Advanced Welding Methods and Equipment

Ruifeng Li · Taotao Li

Advanced Welding Methods and Equipment

Science Press
Beijing

Springer

Ruifeng Li
School of Materials Science
and Technology
Jiangsu University of Science
and Technology
Zhenjiang, Jiangsu, China

Taotao Li
School of Materials Science
and Technology
Jiangsu University of Science
and Technology
Zhenjiang, Jiangsu, China

ISBN 978-981-97-4111-3 ISBN 978-981-97-4109-0 (eBook)
https://doi.org/10.1007/978-981-97-4109-0

Jointly published with Science Press
The print edition is not for sale in China mainland. Customers from China mainland please order the print book from: Science Press.

This Springer imprint is published by the registered company Springer Nature Singapore Pte Ltd.
The registered company address is: 152 Beach Road, #21-01/04 Gateway East, Singapore 189721, Singapore

Preface

As a common basic manufacturing technology, welding has developed from a traditional thermal processing technology to an engineering discipline integrating materials, metallurgy, structure, mechanics, electronics, and others, playing an important role in the modern industrial economy. In recent years, traditional manufacturing technology has gradually failed to meet the needs of the manufacturing industry with the rapid development of the economy. The deep integration of industrialization and informatization, the continuous development of new materials and structures, special working environments, and the need to improve product quality and reliability and reduce production costs and energy consumption have all put forward higher requirements for welding technology. Consequently, welding technology is in urgent need of innovation and welding-related industries also need to be on par with the international advanced level. Advanced welding technologies such as friction stir welding, electron beam welding, and laser welding are becoming more and more mature, and their application in the manufacturing industry is gradually increasing. A better understanding of the various new welding techniques could contribute to the growth of the manufacturing industry and inject fresh blood into advanced manufacturing technology.

Currently, there are more and more research institutes engaged in advanced welding technology, and the number of practitioners has increased exponentially. However, the existing reference materials about advanced welding equipment and technology are mainly based on discrete papers or patents. English textbooks presenting a body of knowledge are still rare, which is incompatible with the current promotion, research, and application of advanced manufacturing technology. Therefore, it is necessary for a specialized book that comprehensively deals with advanced welding equipment and technology to support the development of the discipline. We have spent more than two years referencing a number of original English monographs and textbooks on welding equipment and technology, and this English textbook has been compiled after careful understanding and generalization. In order to enable senior engineering students and postgraduates to be familiar with professional vocabulary as soon as possible, we directly adopt English originals. In the information age where knowledge is being updated more and more rapidly, they are encouraged to extensively read original works in foreign languages reflecting the latest developments in their own fields, so that they can keep abreast of the latest scientific research results. In order to improve the teaching effect of this English textbook, we intentionally attach exercises in each chapter to deepen the understanding and consolidation of related professional knowledge. In addition, this English textbook has been taught as an elective course for postgraduates many times. The contents of the book have been reduced and modified according to the comments and suggestions of the experts concerned.

This book mainly introduces laser welding, laser-arc welding, stir friction welding, narrow gap arc welding, underwater welding, and several more widely applied advanced welding technologies and their principles. And the advantages and defects of different welding methods are analyzed and the development and application in manufacturing industries are specifically given. The authors of this book are all teachers of welding direction at Jiangsu University of Science and Technology. Although much of the content in this book has been published in journals or from other academic monographs, the content is not an overview of domestic and international welding equipment and technology. We try to provide self-contained knowledge and methods to facilitate the reader's learning. To the end, thanks to the main contributors of each chapter: Chaps. 1 and 2, Ruifeng Li; Chap. 3, Taotao Li; Chap. 4, Xiaoqiang Zhang; Chap. 5, Hangyu Yue; Chap. 6, Yue Zhao; and Chap. 7, Xiaolin Bi. We are grateful for the help of graduate students: Mingjun Yan, Zijian Yuan, Yu Fu, Mao Li, Shengli Liu, Xiaolong Xie, and Qiaoqing Jin. In addition, we would like to thank the corporate experts for guidance and advice on this book from Prof. Hongyan Liu of Nanjing Baose Company Ltd. and Prof. Zhigang Liu of China Railway Baoqiao (Yangzhou) Company Limited.

We are indebted to a large number of colleagues from whom we have learned, with whom we have discussed and interacted, or who have provided thoughtful comments on parts of this book. Meanwhile, readers are welcome to write in and give their valuable opinions. Author's contact details: Ruifeng Li, li_ruifeng@just.edu.cn; Taotao Li, litaotao@just.edu.cn.

Zhenjiang, China

Ruifeng Li
Taotao Li

Contents

Chapter 1
Laser Welding

1.1 Principles, Characteristics and Classification of Laser Welding

1.1.1 Principles of Laser Welding

(1) **Principles of laser**

The word laser simply means Light Amplification by Stimulated Emission of Radiation. It gives a basic idea of the principles on which laser devices work, i.e., the generation of a monochromatic, coherent, collimated and intense beam of light. Every device capable of producing a laser radiation, the term laser is associated with that for extension. According to laser principles, all laser devices should be composed of three fundamental parts: an active medium, a pump source, and an optical resonator [1–3]. The working principle of laser is shown in Fig. 1.1.

1. Active Medium: The active medium is the material within the laser that can undergo stimulated emission and generate light. It can be a solid, liquid, or gas, depending on the type of laser. The active medium typically contains atoms, molecules, or ions that are excited to a higher energy state by an external energy source.
2. Pump source: excitation of the laser working substance to pump activated particles from the ground state to higher energy levels for particle number inversion. It can be optical excitation, gas discharge excitation, chemical excitation, nuclear excitation, etc. The choice of the excitation source depends on the characteristics of the working substance. Thus, different working substances often require different pumping sources. For example, for solid-state lasers, pulsed xenon lamps, tungsten iodine lamps and other optical excitation methods are generally used, and for gas lasers, electrical excitation methods are used through the discharge of direct excitation of the working material.

R. Li and T. Li, *Advanced Welding Methods and Equipment*,
https://doi.org/10.1007/978-981-97-4109-0_1

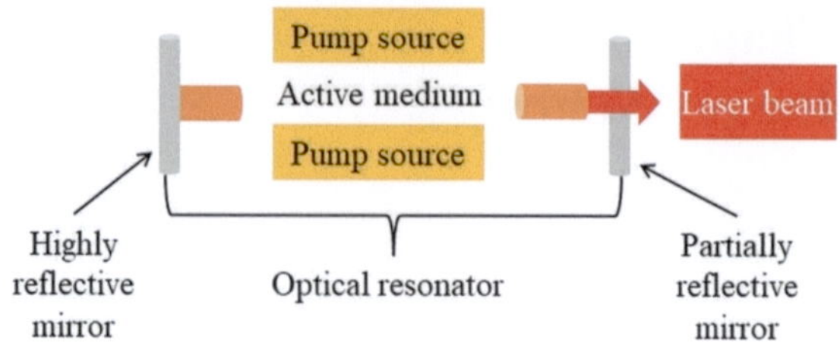

Fig. 1.1 Laser working principle diagram

3. Optical Resonator: The optical resonator consists of two mirrors placed at both ends of the laser. One mirror is fully reflective, while the other is partially reflective. The partially reflective mirror allows a portion of the light to escape, forming the laser beam. The mirrors create an optical feedback mechanism that amplifies the light through multiple reflections, resulting in the generation of a coherent and focused beam of light.

When the active medium is excited by the pumping mechanism, the stimulated emission process occurs. Photons with the same energy and phase as the incident photons are emitted, triggering a chain reaction where additional excited atoms or molecules emit more photons. This process leads to a cascade effect, resulting in a population inversion between the higher and lower energy levels of the active medium. Once the population inversion is achieved, the optical resonator allows the back-and-forth reflection and amplification of photons. As the photons pass through the active medium, they interact with the excited atoms or molecules, stimulating them to emit more photons. This feedback loop causes the amplification and coherence of the light, resulting in the production of a highly intense and focused laser beam.

(2) **Laser beam quality**

Laser welding requires a small beam spot capable of deep penetration, so a high-intensity laser beam with low diffusion is considered high quality. The quality of a laser beam is defined by the convergence of the beam and the intensity of the laser. However, because the laser beans used for laser welding are invisible to the naked eye, beam quality is controlled by visualizing the laser with a measuring device such as a dedicated laser measuring instrument.

The quality of the laser beam is largely determined by two factors: wavelength and energy density (or power density). These factors are largely influenced by two separate factors (beam parameter product (BPP); laser intensity), which can be managed as needed [4].

1. Beam parameter product (BPP)

Beam parameter product (BPP) is the value that defines the influence of a lens on the spread and convergence of the laser beam. This value can be expressed as the product of the radius ω_0 of the beam waist and the full width at half-maximum θ of the divergence angle of the beam. When passing through the converging lens, the following relationship between the spot diameter and the divergence angle after the condensing point of the lens applies.

$$\omega_1\theta_1 = \omega_2\theta_2 = \text{cons t} \quad \text{BPP} = \omega_0 \times \theta = M^2 \times \frac{\lambda}{\pi} \tag{1.1}$$

Here, M^2 describes the beam quality in units of [mm·mrad].

From this it can be inferred that a smaller beam waist and convergence angle and a smaller BPP value will enable deeper penetration and faster processing. In welding and other laser processing, M^2 is often used as a parameter for expressing beam quality. For example, the beam quality for fiber lasers is generally indicated by the M^2 value. At a wavelength of λ, the relation between M^2 and BPP can be expressed using the following equation.

$$\text{BPP} = M^2 \times \frac{\lambda}{\pi} \quad M^2 = \frac{\theta \cdot \omega}{\theta_0\ \omega_0} \tag{1.2}$$

2. Laser intensity

The intensity of a laser is the laser power divided by the convergence, where the convergence is the spot area multiplied by the solid angle of the beam. This means that higher outputs and higher convergences will create higher intensities.

$$B = \frac{P}{\Omega} = \frac{P}{\pi\omega_0^2 \times \pi\theta^2} = \frac{P}{\pi^2\omega_0\left(\frac{\lambda}{\pi\omega_0}\right)^2} \tag{1.3}$$

1.1.2 Characteristics of Laser Welding

(1) Characteristics of laser

1. Directivity

Ordinary light sources emit light in all directions, and the direction of the laser can be limited to several millimeters arc solid angles, which can increase the illumination in the direction of illumination by tens of millions of times. Laser alignment, guidance and ranging take advantage of good directivity.

2. High brightness

Laser is the brightest light source at present, rivaled only by the powerful flash of light at the moment of a hydrogen bomb explosion. The solar light intensity is about 10^3 W/m^2 and a high power laser intensity is 7–14 orders of magnitude higher than that of the solar light. Thus, although the total energy of the laser is not necessarily large, it is easy to produce high pressure and tens of thousands of degrees Celsius or even millions of degrees Celsius at a tiny point due to high concentration of energy. Laser drilling, cutting, welding and laser surgery take advantage of this feature.

3. Mono-chromaticity

Light is an electromagnetic wave. The color of light depends on its wavelength. The light emitted by ordinary light source usually contains various wavelengths, which is a mixture of various colors of light. Sunlight contains infrared light, ultraviolet light and other invisible light. The wavelength of a certain laser is only concentrated in a very narrow spectral band or frequency range. For example, the wavelength of He-Ne laser is 632.8 nm. Because of the good mono-chromaticity of laser, it provides a very favorable means for precision instruments to measure and stimulate some chemical reactions and other scientific experiments.

4. Coherence

Interference is an attribute of fluctuations. Because laser has the characteristics of high directivity and high mono-chromaticity, it must have excellent coherence. This feature of laser makes holography a reality.

(2) **Advantages and disadvantages of laser welding**

Laser beams can be focused to a very small point compared with arcs used in arc welding. Increasing the energy density using a converging lens makes it possible to use laser welding for localized welding and for joining materials with different melting points. This type of welding is also suitable for detailed welding applications thanks to reduced heat effects, a thin bead, and no generation of processing reaction forces. So laser welding has the following characteristics.

1. Advantages of laser welding:

   ① High precision: Laser welding allows for precise control over the heat input and offers a narrow and well-defined heat-affected zone. This precision allows for accurate and consistent welds, particularly in delicate and intricate work-pieces.
   ② Deep penetration: The focused and concentrated laser beam can penetrate deeply into the work-piece, resulting in high-depth-to-width ratio welds. This characteristic is particularly beneficial for welding thick materials or when a strong joint is required.
   ③ Small heat-affected zone: Laser welding generates minimal heat input, leading to a small heat-affected zone. This reduces the risk of distortion and minimizes the need for post-welding processes such as grinding or machining.
   ④ Minimal distortion: Laser welding produces minimal heat-affected zones and minimal distortion in the work-piece. The localized and controlled heating reduces the chances of warping or affecting the surrounding areas of the weld joint.

⑤ Non-contact process: Laser welding is a non-contact process, which means that the laser beam does not physically touch the work-piece. This characteristic eliminates the need for consumables or direct tool-to-work-piece contact, minimizing the chances of contamination or damage.
⑥ High welding speed: Laser welding is known for its high welding speed, enabling efficient and rapid production processes. The concentrated energy of the laser allows for faster weld travel speeds, resulting in shorter cycle times and increased productivity.
⑦ Versatility: Laser welding is applicable to a wide range of materials, including metals, alloys, plastics, and even dissimilar materials. It is commonly used in various industries, such as automotive, aerospace, electronics, and medical, due to its versatility and ability to join different materials efficiently.
⑧ Remote operation: Laser welding can be performed remotely using robotic arms or automated systems. This characteristic enhances worker safety by minimizing direct exposure to high temperatures, fumes, and potential hazards.

2. Disadvantages of laser welding:

① Equipment cost: Laser welding equipment can be expensive to purchase and maintain, making it less accessible for small-scale or budget-constrained operations.
② Limited gap tolerance: Laser welding requires precise fit-up and minimal gaps between the work-piece edges. Large gaps may result in weld defects or incomplete fusion.
③ Sensitivity to surface conditions: Laser welding is sensitive to surface conditions such as dirt, oxidation, or reflective surfaces. Proper surface preparation is essential to achieve optimal weld quality.
④ Power consumption: Laser welding equipment requires a significant amount of electrical power, which can contribute to higher energy costs.
⑤ Safety precautions: Laser welding involves high-intensity laser beams that can pose a risk to human operators. Strict safety protocols and protective measures must be followed to ensure the well-being of personnel.

1.1.3 Classification of Laser Welding

With laser welding, changing the output power and oscillation form of the laser beam is possible, making the method highly adaptable to various applications for use in various assembly processes.

(1) Heat conduction welding and deep penetration welding

Laser welding includes heat conduction welding and keyhole (deep penetration) welding, characterized by variations in the laser beam output is shown in Fig. 1.2. The differences between the two laser welding processes are given in Table 1.1.

In heat conduction welding, the focused laser beam melts the base materials at the joining point. Solidification produces a smooth weld seam that does not usually require any finishing. This is why heat conduction welding is often used for visible seams. Due to the excellent consistency and controllability of the laser power, the process can also be used to join thin-walled components. The energy enters the work

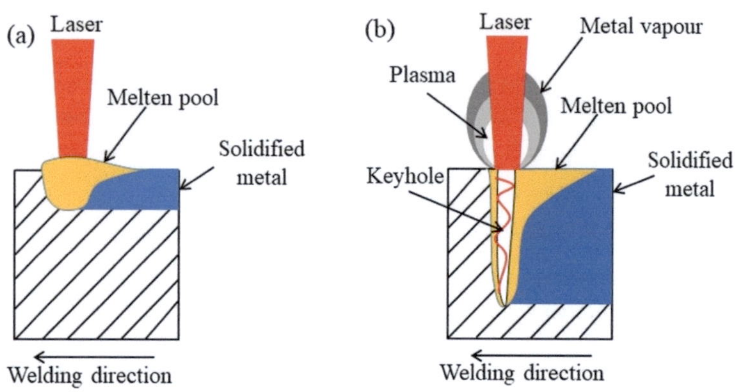

Fig. 1.2 Schematic diagrams of different laser welding methods: **a** heat conduction welding, **b** deep penetration welding

Table 1.1 Differences between heat conduction welding and deep penetration welding

Comparison	Heat conduction welding	Deep penetration welding
Tolerance	Accuracy required on both surfaces	Accuracy required only on one surface
Heat efficiency	Approx. 35%	Approx. 95%
Welding efficiency level	Low	High
Welding speed	Slow	Fast
Energy required for 1 mm seam	High	Low
Heat-based distortion	Slight distortion	Minimal distortion (less than heat conduction welding)
Seam size	Large	Small
Seam smoothness	Smooth	Rough

piece by heat conduction only, which limits the welding depth. Heat conduction welding is a laser welding method that features a low output. This makes for a penetration depth of no more than 1 to 2 mm. With the ability to handle a relatively wide power range, heat conduction welding can be adjusted to the ideal power level, and the shallow penetration makes it possible to weld materials that are susceptible to heat effects under optimal conditions. This welding type is used for butt joints, lap joints, and other welding applications for thin plates, and can also be used for welding hermetic seals and other seals. Heat conduction welding is also suitable for volatile alloys such as magnesium and zinc, for which keyhole welding is not suitable [5].

Deep penetration welding requires very high power densities of about 1 MW/cm^2. The laser beam then not only melts the metal, but vaporizes it. The vapor pressure creates a narrow vapor-filled hole called a vapor capillary or keyhole. As the laser beam advances through the work piece, the molten metal flows around the capillary and solidifies in its trail. This produces a narrow, deep weld with a uniform structure. Deep penetration welding is distinguished by a great welding depth and fast process speed. Due to the high welding speed, the heat-affected zone is narrow and distortion is minimal. Deep penetration welding uses a high-output laser for high-speed welding. The narrow, deep penetration allows for uniform welding of internal structures. Because the heat-affected zone is small, distortion of the base material due to welding can be minimized. This method is suitable for applications requiring deep penetration or when welding multiple base materials stacked together (including for butts, corners, laps and flange joints).

(2) **Pulsed laser beam welding and Continuous laser beam welding**

The oscillation form of a laser can be either pulsed or continuous. Pulse oscillation lasers generate a pulse wave by controlling (through light modulation) the intensity, wavelength, and phase of the laser beam. Methods for generating a pulse also include controlling the Q value of the laser oscillator as with Q-switch lasers. The "Q value" is the ratio of energy consumed in each cycle to the accumulated energy. The shape of the weld bead also varies for each oscillation form.

Pulsed laser beam welding and continuous laser beam welding, characterized by different laser oscillation modes are shown in Fig. 1.3. The pulsed laser beam with a short wavelength, powerful laser is ideal for spot welding, while continuous laser beams are ideal for seam welding. Both welding methods are also suitable for welding materials with a high melting point or with high thermal conductivity due to a very low thermal effect during welding. When welding with a laser, the shape and quality of the weld seam can be varied and thus optionally adapted to the task.

1.2 Equipment of Laser Welding

Laser beams can be extracted by using a gas, solid, or liquid as a medium. Of these methods, welding lasers use a gas or solid as the medium. Therefore, the types of lasers used for laser welding can be generally divided into gas lasers or solid-state

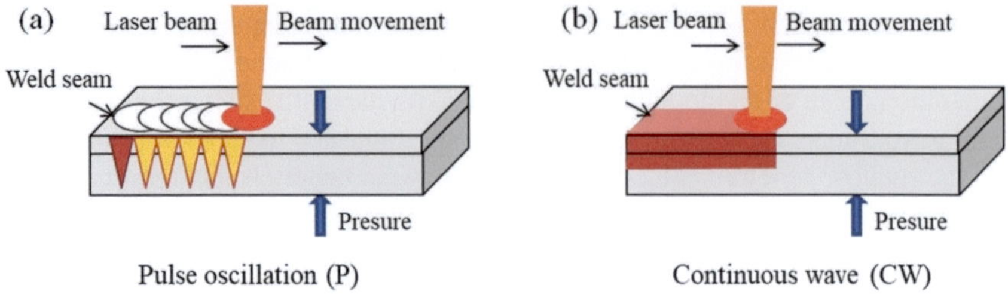

Fig. 1.3 Schematic diagrams of different laser welding methods: **a** pulsed laser beam welding, **b** continuous laser beam welding

lasers. The laser for welding requires high power density (10^{10}–10^{12} W/cm^2), and the power density distribution is fundamental mode with good beam quality. Gas laser welding uses carbon dioxide (CO_2) or other gases to produce light. Solid-state laser welding uses ores such as yttrium, aluminum, and garnet (as with YAG laser welding) to produce light. Recently, fiber laser welding using optical fibers for oscillation, as well as semiconductor lasers and disk lasers developed from YAG lasers have increased in popularity. The main parameters of different laser welding equipment are given in Table 1.2.

(1) **Solid-state laser**

The schematic of a solid-state laser is shown in Fig. 1.4. A solid-state laser uses ore such as yttrium, aluminum, and garnet (YAG) or yttrium vanadate crystal (YVO4) as the laser medium. Because solid-state lasers have a large laser output per unit volume, a large laser output is possible even with a small resonator.

(2) **Gas laser**

The schematic of a gas laser is shown in Fig. 1.5. A gas laser uses a gas such as CO_2 as the laser medium. Compared with solid-state lasers or other laser mediums, a gas medium is more uniform with less loss, and the resonator can be larger to allow for larger laser output.

Table 1.2 Main parameters of different laser welding equipment

Laser welding equipment	Oscillation form	Output	Conversion efficiency (%)
Gas (CO_2 laser)	P/CW	P: 10 MW CW: 20 kW	Max. 20
Solid-state laser (YAG)	P/CW	P: 10 MW CW: 400 W Q-switch: 10 MW	Max. 3
Semiconductor lasers	P/CW	P: 10 W CW: 100 MW	Max. 100

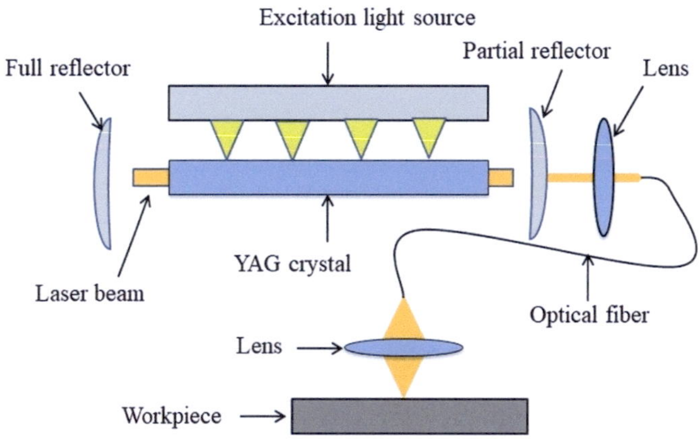

Fig. 1.4 Schematic diagram of a solid-state laser

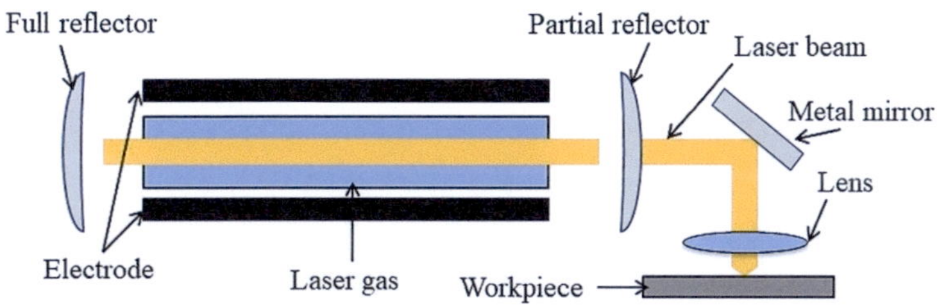

Fig. 1.5 Schematic diagram of a gas laser

(3) **Semiconductor lasers**

Semiconductor lasers use III-V semiconductors or IV-VI semiconductors for the laser medium. Both semiconductor types ensure device compactness while also offering a large laser output. Although semiconductors are solid objects, they are sometimes classified separately from solid-state lasers in laser technology fields.

1.2.1 CO_2 Laser

The schematic of a CO_2 laser is shown in Fig. 1.6. A CO_2 laser is a device that utilizes carbon dioxide as the gain medium and nitrogen (N_2), helium (He). To some extent, it also uses hydrogen (H_2), water vapor, oxygen and/or xenon (Xe) to improve the laser's effectiveness for light application, by emission of radiation that is stimulated. The laser is pumped electrically via an electrical gas discharge. The electrical gas discharge can be operated using AC current, DC current or in the radio frequency domain. It operates at a wavelength of 10 ~ 6 μm.

The construction of CO_2 lasers involves:

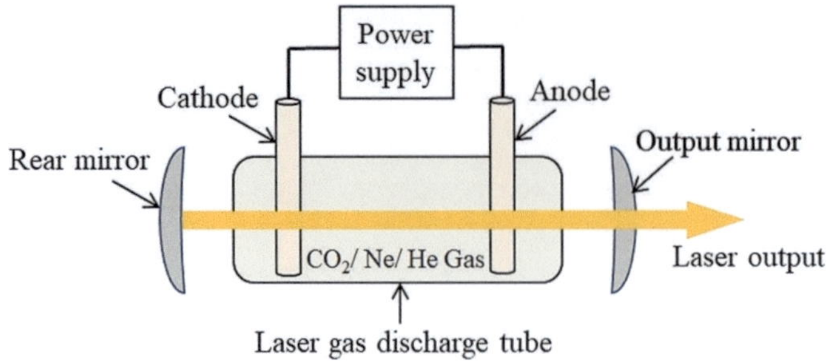

Fig. 1.6 Schematic diagram of a CO_2 laser

(1) Modified materials are needed in the construction of CO_2 lasers because they operate in the infrared region of the spectra. The mirrors are coated with silver and the windows and lenses are made of germanium or zinc selenite. For high power applications, gold mirrors, zinc selenite windows and lenses are used. Others use diamond windows and lenses.
(2) A CO_2 laser consists of a quartz discharge tube with a diameter of 2.5 cm and a length of about 5 m. The discharge tube contains a mixture of CO_2, N_2 and He in a ratio of 1:2:3 with water vapor. Transitions between vibrational states of CO_2 molecules and rotational levels of CO_2 molecules produce a laser action. The construction of a CO_2 laser is simple and the output of this laser is continuous.
(3) An electrical discharge is used as a means of pumping and achieving population inversion. Electrons collide with carbon dioxide molecules and pump them to excited states. During an electric discharge inside the machine, electrons and nitrogen molecules collide and the electrons are raised to their excited state. N_2 molecules in the excited state then collide with CO_2 atoms in the ground state and are excited to higher electronic, vibrational, and rotational levels. Because of the closeness of the excited level of nitrogen and the E5 level of the CO_2 atom, a population increase in the E5 level occurs. The laser action is triggered by any of the spontaneously emitted photons in the tube. This occurs when the population inversion is reached.

The characteristics of CO_2 lasers:

(1) It is a molecular gas laser type.
(2) Active medium: A mixture of CO_2, N_2 and He or water vapor is used as an active medium.
(3) The pumping method used is the electrical discharge method.
(4) It consists of two concave mirrors forming a resonant cavity.
(5) A continuous or pulse wave may be the output.
(6) The output of the wavelength is 10.6 μm.

The gas mixture of N_2, He, and CO_2 is the active medium. The transition of the laser occurs between the vibrational states of the molecules of CO_2 a laser beam has the power to concentrate a huge amount of power on a small area. As a result, it is used in the industry to weld, cut, and make holes. Industrial application of CO_2 lasers:

(1) CO_2 laser welding. During welding, two components are joined by melting the specific materials. High temperatures are needed in the process of melting and then joining the material, but temperature must not be too high to evaporate the material. Two dissimilar metals can be joined using a CO_2 laser. CO_2 laser welding is used in the manufacture of aircrafts and automobiles. Laser welding is also used in electronics.

(2) CO_2 laser cutting. The CO_2 laser cutting processing technique for sheets utilizes a gas laser that is electrically driven. A CO_2 laser is used to remove part of a material from a substance. Materials cut can be metal or nonmetal such as titanium, stainless steel, ceramic glass, plastic and wood.

Benefits and drawbacks of CO_2 lasers: It has simple construction, continuous output of the laser, high efficiency and high output power. The gas tube length can be extended to increase the output. However, the action of the laser can get affected by the contamination of oxygen by carbon monoxide. The operating temperature is a major contributor in the determination of the output power of the laser. Damage of eyes because of the invisible (infra-red region) can occur to our eyes.

1.2.2 *Nd: YAG Laser*

The schematic of an Nd: YAG laser is shown in Fig. 1.7. Neodymium-doped Yttrium Aluminum Garnet (Nd: YAG) laser is a solid state laser in which Nd: YAG is used as a laser medium. These lasers have many different applications in the medical and scientific fields for processes. Nd: YAG laser generates laser light commonly in the near-infrared region of the spectrum at 1064 nm. It also emits laser light at several different wavelengths including 1440, 1320, 1120, and 940 nm.

Nd: YAG laser consists of three important elements: the energy source, the active medium and the optical resonator. ① The energy source or pump source supplies energy to the active medium to achieve population inversion. In Nd: YAG laser, light

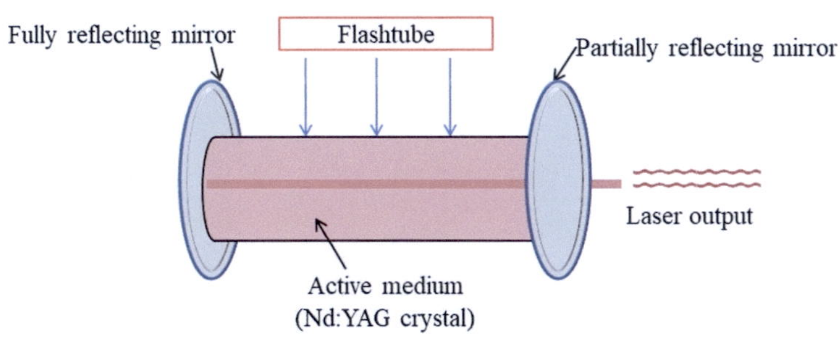

Fig. 1.7 Schematic diagram of an Nd: YAG laser

energy sources such as flashtube or laser diodes are used as energy sources to supply energy to the active medium. ② The active medium or laser medium of the Nd: YAG laser is made up of a synthetic crystalline material (Yttrium Aluminum Garnet (YAG)) doped with a chemical element (neodymium (Nd)). The lower energy state electrons of the neodymium ions are excited to the higher energy state to provide lasing action in the active medium. ③ The Nd: YAG crystal is placed between two mirrors. These two mirrors are optically coated or silvered. Each mirror is silvered or coated differently. One mirror is fully silvered whereas, another mirror is partially silvered. The mirror, which is fully silvered, will completely reflect the light and is known as a fully reflecting mirror. On the other hand, the mirror which is partially silvered will reflect most of the light but allows a small portion of light through it to produce the laser beam. This mirror is known as a partially reflecting mirror.

Advantages of Nd: YAG laser:

(1) Nd: YAG laser has low power consumption.
(2) Nd: YAG laser offers high gain.
(3) Nd: YAG laser has good thermal properties.
(4) Nd: YAG laser has good mechanical properties.
(5) The efficiency of Nd: YAG laser is very high as compared to the ruby laser.

1.2.3 Disk Laser

The schematic of a disk laser is shown in Fig. 1.8. Disk laser is different from the traditional solid-state laser in the shape of the laser working substance. The innovative idea of changing the rod-like crystal of the traditional solid-state laser to the disk crystal has pushed the solid-state laser to a new era. Disk laser has been widely used in the manufacturing industry with its excellent beam quality and conversion efficiency.

An important problem of laser is the cooling of laser working material, and the cooling effect is directly related to the quality of laser. Because the traditional rod-like Nd: YAG laser crystal can only be cooled side by side, i.e., the cooling must be realized by the radial heat conduction of the crystal rod, the temperature in the rod is parabolic distribution, resulting in the formation of so-called thermal lens in

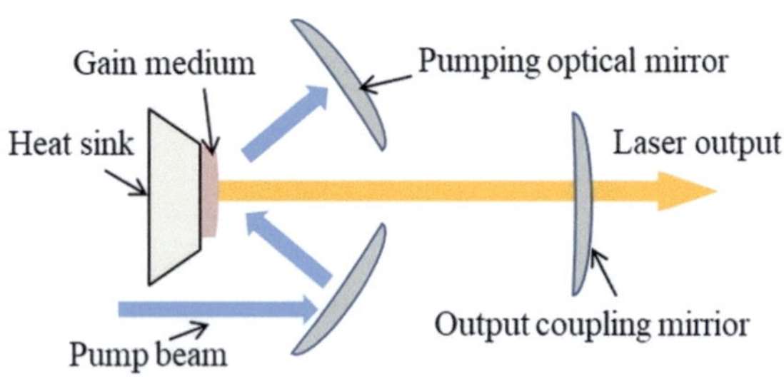

Fig. 1.8 Schematic diagram of a disk laser

the rod. This thermal lens effect will seriously affect the quality of the laser beam and change with the change of the pump power. The larger the pump power is, the larger the thermal lens effect is, and the shorter the focal length of the thermal lens is. The laser may even change from a steady state to an unsteady state, which seriously limits the development of the solid-state laser in the direction of high power.

Disk laser can maintain good beam quality even at high power, because it solves the thermal lens effect problem of traditional solid laser. The characteristics of high conversion efficiency and low operation cost make it play a unique advantage in industrial applications.

1.2.4 Semiconductor Laser

The schematic of a semiconductor laser is shown in Fig. 1.9. Also called laser diode, semiconductor (LD) laser is a technique that uses a laser beam generated by an electric current passing through a semiconductor as the heat source. Because the lamp is not used as the excitation source, devices can be compact, and maintenance such as lamp replacement is not necessary. These advantages have made this technique one that is expected to become widespread in the future. CO_2 lasers and YAG lasers create an oscillated beam by supplying (exciting) the energy required for laser emission to a medium (CO_2 gas or YAG). A semiconductor laser, however, creates an oscillated beam as an electric current passes through the semiconductor. The light emission mechanism is the same as a light-emitting diode (LED) with the light being emitted as the current passes in the forward direction of the P-N junction. The high-quality laser beam generated with high conversion efficiency is oscillated to output of a high-energy beam comparable to an electron beam.

The necessary conditions for laser generation by semiconductor lasers mainly include five aspects: pump source, working substance (particle number inversion), resonant cavity, positive feedback and output.

Semiconductor materials are used as working materials in semiconductor lasers. Common materials include GaAs, CdS, InP, ZnS etc. Semiconductor laser devices can generally be divided into homo-junction, single hetero-junction and double

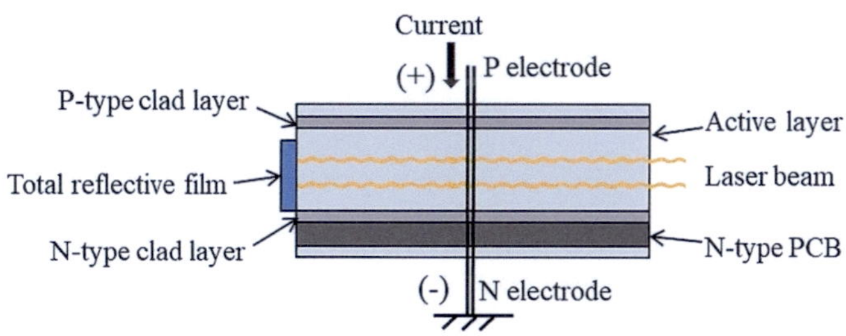

Fig. 1.9 Schematic diagram of a semiconductor laser

hetero-junction. Homo-junction lasers and single hetero-junction lasers are mostly pulsed devices at room temperature, while double hetero-junction lasers can work continuously at room temperature.

The excitation modes include electrical injection, electron beam excitation and optical pump excitation. Electrically implanted semiconductor lasers are generally made of GaAs, InAs, InSb and other materials. The pump source is provided by the positive bias at both ends. That is, the current is injected along the positive bias to excite, and the stimulated emission is generated in the junction plane region. Among semiconductor laser devices, GaAlAs-GaAs diode laser with double hetero-structure has good performance and wide application.

In order to make the semiconductor laser generate laser light, attenuation must also be considered, that is, only when the gain is equal to or greater than attenuation, the laser can output laser light. The attenuation of laser mainly includes the decrease of carrier number due to stimulated radiation (i.e. the number of excited particles), the decrease of carrier number due to a small amount of spontaneous radiation and the decrease of carrier number due to nonradioactive collision with the medium. Therefore, the general semiconductor laser needs a certain amount of injected current to emit laser, which is called threshold current.

1.2.5 Fiber Laser

The schematic of a fiber laser is shown in Fig. 1.10. A fiber laser is a type of solid-state laser that uses an optical fiber as the medium. The amplification fiber is an optical fiber called a “double-clad fiber” with a core that is doped with a rare earth element. The output side is also equipped with a low-reflectance mirror, and the incident side is equipped with a high-reflectance mirror.

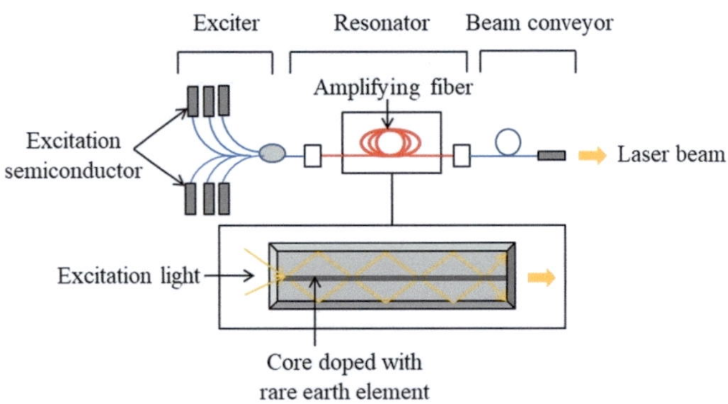

Fig. 1.10 Schematic diagram of a fiber laser

With fiber laser oscillation, the excitation light generated from the exciter first enters the amplifying fiber and propagates through reflections inside the amplifying fiber. The rare-earth element doped in the core of the amplifying fiber absorbs the excitation light, and electron population inversion occurs between the ground level and the metastable level to emit light. The emitted light is repeatedly reflected between the high-reflectance mirror and the low-reflectance mirror in the resonance section before being oscillated as a laser beam.

Fiber laser has the following characteristics: no adjustment, no maintenance, high stability, high electro-optic conversion efficiency; small size, optical fiber transmission, and excellent beam quality. The beam quality of single-mode fiber laser is close to the theoretical limit, and that of multimode fiber laser is close to the TEMO mode of CO_2 laser. The excellent beam quality makes the fiber laser have the application ability in many fields, such as welding, cutting, drilling and deposition. Compared with the traditional laser, the laser with the same power can obtain greater penetration, faster welding speed, smaller welding deformation and higher welding quality.

As a new generation of laser source, fiber laser is easier to transmit energy than CO_2 gas laser and simplifies the system design. After parameter optimization, the welding and cutting quality of stainless steel and aluminum alloy can achieve excellent cutting effect of gas laser.

1.3 Laser Welding of Metallic Materials

1.3.1 Laser Welding of Homogeneous Metals

(1) **Laser welding of steel**

Steel is classified into various categories based on its composition and properties. The main types of steel include carbon steel, alloy steel, stainless steel, and tool steel.

Carbon steel is primarily made up of iron and carbon, with low amounts of other elements. It is known for its strength, durability, and affordability. Carbon steel is commonly used in construction, automotive, and machinery industries.

Alloy steel is made by adding various elements such as manganese, nickel, or chromium to carbon steel. This enhances its strength, hardness, and resistance to wear and corrosion. Alloy steel is frequently used in the production of aircraft, pipelines, and high-stress components.

Stainless steel contains chromium in addition to iron and carbon. This alloy provides excellent corrosion resistance and a lustrous appearance. Stainless steel is widely utilized in kitchen appliances, medical equipment, and architectural structures.

Tool steel is specifically designed for manufacturing tools. It is known for its high hardness, resistance to heat, and wear resistance. Tool steel is commonly used in the production of cutting tools, molds, and dies.

Overall, these different classifications of steel offer a range of properties and applications, catering to diverse industrial needs.

Laser welding of steel offers several advantages and finds application in various fields. One major advantage is the precise control and concentrated heat input provided by the laser beam, leading to minimal distortion of the welded material. Additionally, laser welding offers high welding speeds, resulting in increased productivity. The ability to achieve deep penetration and narrow weld zones makes laser welding suitable for joining thin materials with high welding strength. It also allows for the welding of complex geometries and hard-to-reach areas.

In terms of applications, laser welding is widely utilized in the automotive industry for joining body parts, exhaust systems, and engine components. It is also used in aerospace applications for welding aircraft structures and engine parts. Furthermore, laser welding is employed in the production of appliances, electronics, and various metal work pieces where precise and high-quality welds are required. The advantages of laser welding in terms of precision, speed, and versatility make it a preferred choice for joining steel in numerous industries, ensuring strong and reliable welds.

1. Laser welding of stainless steel

Research has shown that deep penetration autogenously laser welding of 304L plates is feasible [6]. Buddu et al. [7] have performed 8 mm thick welds using a 3.5 kW CO_2 laser. Zhang et al. [8] have achieved 12 mm thick welds applying a 10 kW fiber laser. The penetration of the melt runs was measured from etched cross-sections. The penetration increased with laser power and decreased with welding velocity, in line with an increase in heat input [9]. High powered diode lasers are capable of deep penetration welding applications. For welding of plates in the 10 mm thickness range, the more economical diode laser systems may become a competitor to other systems with high quality laser beams which may be more expensive in both procurement and operation.

2. Laser welding of thick-section steel

Since its invention, steel has been the predominant engineering material for structural construction applications, with various grades and thicknesses available. Thick section steel has the characteristics of high strength, good toughness, and low cost. It's typically used as the main body of large-scale structures, has been widely employed with increasing demand in industries such as oil and gas pipeline, power plants, construction engineering, shipbuilding, and ocean engineering [10–14]. In the manufacturing of large-scale structures, the welding technique selected has a significant impact on safety, design flexibility, environmental considerations, and economic factors. High-power laser welding is a favored option for large thick-section components due to its high energy density, fast welding speed, deep penetration, low distortion, and ability to produce high-quality joint. The most noteworthy advantage is that high-power laser welding can significantly improve efficiency by reducing the number of weld layers required, sometimes even to a single pass. To the best of the researchers' knowledge, in the laser power range up to 20 kW, the penetration is approximately 1 mm per 1 kW, while it drops to approximately 0.5 mm per 1 kW from 20 to 100 kW [15, 16]. Hence, thick-section steel and high-power laser welding are naturally linked together.

(2) **Laser welding of aluminum alloys**

Aluminum and aluminum alloys play an important role in e-mobility because of the ideal combination of lightness and strength. In welding and joining processes, aluminum has high demands on the processes used. With comprehensive process know-how and the right products, we make you fit for the technological transformation towards e-mobility.

Aluminum alloys are more widely used nowadays in a vast variety of industries due to their low weight (density 2.7 g/cm^3), high corrosion resistance, excellent formability, high toughness at cryogenic temperatures, face-centered cubic (FCC) crystal structure, high thermal and electrical conductivity, and they are relatively inexpensive. In addition, Al has excellent recyclability [17], and is thus widely used in the production of car bodies [18], different structural components, shipbuilding, and as packaging material. Recently, aluminum alloys have become more attractive in electrical battery assemblies [19] due to high conductivity and low density. Aluminum alloys exhibit good mechanical properties among other materials and possess high strength-to-weight ratios [20]. Their weld-ability may have significant challenges, especially in joining precipitation-hardened aluminum alloys with high requirements for weld quality.

Laser beam welding is commonly carried out without filler wire and is more frequently used for joining thin sheets (< 3.0 mm), typically applied in the automotive and aerospace industries. The addition of filler wire can be complicated and requires tedious optimization. The laser welding of thicker plates (> 5 mm) is an attractive option due to the enhanced productivity.

With the rapid development of ultra-high power fiber laser, high power laser are widely applied in industry. Systematic studies on the effect of parameters on ultra-high laser power (over 20 kW) during welding under atmospheric and sub-atmospheric pressures were conducted [21]. A 130 mm thick 5A06 aluminum alloy butt joint was double-side welded under sub-atmospheric pressure. When the ambient pressure was below 10^2 Pa, the porosity in the joints was invisible.

(3) **Laser welding of titanium alloys**

The microstructure of pure Ti is α phase with HCP (hexagonally close-packed) and once heated above 882 °C, it undergoes allotropic phase transformation into the ductile body centered cubic (BCC) β phase comprising of a greater number of slip systems. This gives Ti a major advantage because the properties of Ti are highly dependent on the heat treatment and subsequently are affected by the heating and cooling cycles imparted during laser welding. The alloying of Ti can enhance the secondary properties, thermo-mechanical processing, strengthening via heat treatment and microstructural changes, etc. [22]. Ti can be alloyed based on the ability of the alloying element to stabilize the primary α or β phase depending on the atomic radius. For instance, solute elements of Al with an atomic radius in the range of 0.085–0.115 nm diffuse substitution ally by replacing one of the bulk atoms in the lattice.

Whereas, atoms with radii lower than 0.059 nm occupy interstitial sites between the larger solvent atoms. Stabilizing elements are added to prevent subsequent phase changes, increase or decrease the β-transits temperature range and enhance mechanical properties through heat treatment. The α phase is stabilized by various elements such as Al, O, N, Ga, etc., while the β phase can be stabilized by Mo, V, W and Ta. Accordingly, the Ti-alloys are classified as α or near α, α + β, β or near β alloys.

Laser welding of Ti and its alloys has been carried out in similar and in dissimilar material combinations. However, to reduce the scope of this review, limited topics for similar welding of Ti alloys have been considered. In general, efforts have been undertaken by researchers to investigate and optimize the effect of processing parameters on different types of Ti alloys. Fundamental studies that are essential to examine the keyhole formation mechanism, porosity development, heat flow and geometrical features of the weld have been accompanied by both experimental and simulated formulations. Similarly, dissimilar welding studies have been dissected based on different techniques employed to enhance the joint strength and suppress the intermetallic compounds (IMC) formation. Hence, researchers have focused on an alternative modification of laser processing methods like laser offsetting, welding brazing and hybrid welding techniques and have made changes in the materials system by either employing filler materials or by adding single or multiple interlayers. Some studies have also concentrated on investigating the effect of joint type, groove shape and groove angle on the joint efficiency. Different characterization techniques and standards used to evaluate the mechanical performance along with the general applications of Ti-based laser welds are summarized systematically in the following sections. This is accompanied by a detailed review of dissimilar welding of Ti with its dissimilar counterparts.

From the perspective of titanium alloys, its high reflectivity and thermal conductivity also impose higher requirements on laser welding technology. In titanium alloy laser welding, it is necessary to adjust parameters such as laser output energy, spot size, and welding speed to control and adapt to the characteristics of titanium alloys. Furthermore, the oxidation issue of titanium alloys needs special attention, which can be addressed by inert gas protection or the use of shielding agents to prevent oxidation reactions.

Titanium alloy laser welding technology has shown great potential in a wide range of applications. In the aerospace industry, it can be used to manufacture aerospace engine components, aircraft structural parts, and more. Its high-precision welding ability ensures the dimensional and geometric accuracy of the parts, improving product quality and reliability. In addition, titanium alloy laser welding technology has also been widely used in fields such as automotive, medical devices, and energy. In the medical field, it can be used to manufacture biomedical materials such as artificial joints and dental implants, overcoming the thermal effects and degradation issues associated with traditional welding methods, thus providing high-quality connection methods [23].

Despite the many advantages of titanium alloy laser welding technology, it still faces some challenges in practical applications. Firstly, the complexity of laser systems and equipment, as well as high equipment costs and technical requirements, limit the widespread adoption of this technology. Secondly, process control of titanium alloy laser welding requires operators with high skills to accurately adjust the laser welding parameters, thereby improving the quality and efficiency of welding.

In conclusion, titanium alloy laser welding technology overcomes the challenges encountered with traditional welding methods in titanium alloy welding by precisely focusing laser energy on the welding area. Titanium alloy laser welding technology provides a good solution for the connection and processing of titanium alloy materials with its features of high precision, high efficiency and no damage, and has a wide range of application prospects in many fields. Through continuous research and innovation, the titanium alloy laser welding technology will be further improved and promoted.

1.3.2 Laser Welding of Dissimilar Metals

The ability to manufacture a product using a number of different metals and alloys greatly increases flexibility in design and production. Properties such as heat, wear and corrosion resistance can be optimized, and benefits in terms of production economics are often gained. Joining of dissimilar metal combinations is, however, a challenging task owing to the large differences in physical and chemical properties which may be present. Laser welding, a high power density but low energy-input process, provides solutions to a number of problems commonly encountered with conventional joining techniques. Accurate positioning of the weld bead, rapid heating and cooling, low distortion, process flexibility, and opportunities for product redesign are its principal characteristics. The review describes the principles underlying laser welding of dissimilar metal combinations and highlights the above benefits in a number of practical applications. It is concluded that there is potential for its application in many industrial sectors [24].

(1) **Laser welding of steel and aluminum dissimilar metals**

Lightweight aluminum alloy and steel hybrid structures can significantly reduce fuel consumption and decrease air pollution in auto industries [25]. However, the physical properties of steel and aluminum vary greatly, and Fe-Al intermetallic compounds are easily generated during welding, which reduces the mechanical properties of welded joints. Therefore, steel and aluminum dissimilar metal welding has always been a challenging thing. At present, there are various welding methods for steel and aluminum dissimilar metal welding. During the welding of dissimilar metals, laser welding can effectively improve the flexibility and adaptability of welded joints. Therefore, laser welding is considered a viable method for dissimilar metals' welding and is widely used in various industrial fields [26]. In many papers, transition layer and filler wire have been used to improve the tensile of steel-aluminum welded joints, and good results have been obtained. Elementary substances are the most easily considered and exploited in all the transition layers. The elementary substance is added in the form of a powder, foil, coating between the steel and Al alloy surfaces for butt/lap welding.

Good powder fluidity can improve the fluidity of the molten pool. But the amount of powder layer added in the experiment is difficult to control, which is easy to cause the instability of welding results. The thickness of the foil is easy to control. Adding Ti foil is beneficial to reduce the tendency of crack generation. Adding Ni foil can decrease the crack at the fracture surface of the aluminum side, which means that the joint with Ni-foil has crack resistance. The crack resistance also indicates that the joint with the Ni-foil has higher toughness than the joint without the Ni-foil. Among many elementary substances, it is crucial to select an element that can significantly improve the mechanical properties of steel-aluminum welded joints. Ni, Pb, B, Sn elements can well improve the mechanical properties of the welded joint. However, the experimental results of the steel-aluminum dissimilar metal laser welding are influenced by many factors, such as laser process parameters [27, 28], steel and aluminum-based material, thickness of transition layer, lap/butt joints. So the experimental results in different papers cannot be directly compared.

Fill wires are used in the welding process to connect steel and Al alloy. At present, the types of those used filler wires are mainly Al-based filler wire, Sn-based filler wire, Zn-based filler wire. The melting temperature of these filler wires is close to aluminum-based material lower than steel, so it can avoid the melting of the steel and reduce the production of Fe-Al. Therefore, the filler wire generally forms a brazing interface with the steel and a fusion welding interface with the aluminum-based material [29].

(2) **Laser welding of aluminum and titanium dissimilar metals**

The second most common pair of metals welded by a laser beam is titanium and aluminum. Obtaining a durable fixed joint of this pair of metals has been an urgent task since the beginning of spaceship construction and modern aviation. At that time, attempts at heat-press welding of titanium and aluminum adapters were made; at present, other types of mechanical and thermo-mechanical welding classes are actively used. The resulting welded joints were used to manufacture various aircraft systems that passed the tests successfully.

When welding titanium and aluminum, usage of a laser beam has advantages compared to other sources of energy and melting. Because of the high degree of parameters control of laser welding modes, such as power and welding speed, it is possible to precisely control the melting volume of metals and, accordingly, their metallurgical interaction. This can be used to control the volume of IMC formation and composition. Similar to welding of steel and aluminum, when welding titanium and aluminum, butt and lap joints are most common.

Some physical properties of aluminum and titanium differ significantly, for example, the thermal conductivity coefficient (Al-238 W/(m $\times$ K), Ti-22 W/(m $\times$ K)). When butt welding, the basic technique is to offset the laser beam onto one of the alloys. By offsetting the laser beam on aluminum, it is possible to obtain the minimum IMC and maximum mechanical properties. The mechanical properties of welded joints, obtained by offsetting the laser beam on a titanium alloy, are 40–50%

lower than when offsetting the laser beam on an aluminum alloy. This significant difference is explained by the fact that when offsetting the laser beam on an aluminum alloy, the TiAl and $TiAl_3$ IMCs are formed, and when offsetting the laser beam on a titanium alloy, more brittle Al_3Ti, Al_2Ti IMCs are formed. High welding speeds (200–300 mm/s) and a decrease in heat input contribute to the minimization of the IMC thickness. When lap welding, it is reasonable to impact from the side of titanium since molten titanium better interacts with solid aluminum than molten aluminum with solid titanium [23].

(3) **Laser welding of aluminum and copper dissimilar metals**

Aluminum-copper dissimilar metal joints exhibit extensive potential applications in fields such as electric engineering, chemical engineering and refrigeration industry due to their excellent mechanical properties and high electrical and thermal conductivity. For example, using Al-Cu composite structures in power electronic and refrigerator radiators reduces consumption costs and creates a light-weight structure. With the development of new dissimilar alloy welding techniques, the use of Al/Cu joints is receiving increasing attention worldwide.

Currently, aluminum and copper are generally joined by means of fusion welding, braze welding and pressure welding. However, when performing fusion welding, incomplete fusion and impurities can easily be caused due to the large differences between Cu and Al melting points and brittle intermetallic compounds with large electrical resistances may form at the interface between aluminum and copper. Moreover, the linear expansion coefficient of Cu is 1.5 times larger than that of Al, and cracks are likely to be generated during the solidification process, adding to the difficulties associated with fusion welding of Al-Cu joints. In braze welding, several brittle phases, such as $CuAl_2$, Cu_2Al and Cu_3Al_2 phases, are easily produced in the interface of the two metals. These phases reduce the strength and plasticity of the brazing seams. Additionally, in order to removal the dense aluminum oxide film, some corrosive brazing fluxes must be used. The absorption of water by the brazing fluxes causes severe corrosion in the brazing seams, as well as producing harmful environmental pollution. In addition, specialized welding equipment and complex technology in pressure welding of Al-Cu joints limited its applications. Laser welding has several advantages over traditional welding methods, such as low heat inputs, precision-controlled energy inputs and action sites. As a result, the effects induced by the differing physical properties of the metals and the intermetallic components can be reduced or even eliminated by using laser welding, making laser welding become a feasible means for Al-Cu joining. In general, the regulation of the contents of intermetallic components is a key factor in laser welding of Al-Cu joints.

To eliminate the formation of IMC, one of the methods is laser beam offset. When butt welding, the laser beam is offset on aluminum, which allows obtaining the minimal IMC with satisfactory electrical properties. When lap welding, the laser beam impact from the side of the aluminum is reasonable. It is possible to obtain welded joints of aluminum and copper with the tensile shear strength of 95–128 MPa using this technique. In this case, the width of the weld pool can be increased by defocusing or transverse oscillation of the laser beam to increase the electrical conductivity over the contact area of aluminum and copper [30, 31].

The other method is the choice of filler material and the study of the IMC composition, the nature, and conditions of the IMC formation. Since the speeds of laser welding are sufficiently high (up to 200 mm/s), the use of wire filler material for butt welding of small thicknesses (up to 2 mm) will be difficult because the wire will not always have time to melt. In the case of an overlap joint, it is impossible to use the wire as a filler material, therefore the use of thin plates or electroplating of thin layers (10–50 μm) of metals (Cu, Ni, Ag) on one of the welded dissimilar metals will be promising [32, 33].

1.4 Application of Laser Welding

Compared to arc welding and electron beam welding, laser welding is easier to combine with computer control and CAD/CAM. In addition to being easy to incorporate into production lines, laser welding is also suitable for use with robots. Laser welding is a high-accuracy, high-quality, high-speed, high-productivity joining method that utilizes the features of high energy density and small spots for the heat source.

1.4.1 Application of Laser Welding in Shipbuilding

In shipbuilding, the materials processed are mostly plates ranging from 3 to 12 mm thick. The main problem faced by traditional shipbuilding methods is the distortion and deformation caused by the large heat input during the welding process of these plate welding components. During the construction of a ship's hull, approximately 25% of the workload involves shaping and hammering the ship plates to achieve the required flatness and curvature, allowing the plates to be firmly fixed together. One of the major advantages of laser welding technology in shipbuilding is its low distortion. Currently, laser welding is widely accepted in the shipbuilding industry for structural components such as sandwich panels, angle profiles, T-profiles, cross profiles, and butt-welded plates.

Sandwich panels are composite structural panels made up of two outer layers of material (usually metal) with a core material sandwiched in between. Due to their superior structural characteristics, sandwich panels have been widely applied. With the development of high-strength-to-weight ratio metal materials, the combination of sandwich structures and lightweight alloys has reduced overall structural weight and achieved multi-functionality. The combination of new structures and materials offers advantages such as high strength-to-weight ratio, high energy absorption, corrosion resistance, sound insulation, and thermal insulation, making it a cutting-edge project in the early twenty-first century. It's a typical sandwich panels can be classified into various types based on their core structure and shape, such as X-Core, O-Core, I-Core, Z-Core, and V-Core. Due to its various advantages, research on the laser welding of sandwich panels has been conducted since the late twentieth century and has a history of many years.

Laser welding with multi-kilowatt fiber lasers is fast becoming a highly advantageous joining technology in manufacturing industries such as shipbuilding, where it saves production time and cost compared to conventional arc based welding processes. It can be seen in Fig. 1.11 that the penetration depth (measured from top of the weld) and the length of the joint fusion along the intersection decreased as the inclination angle increased. It can also be noticed that the penetration depth and area of the fusion zone correlate with the energy density of the beam. A full penetration weld was obtained only with the 200 μm process fiber and 6° inclination angle [34].

Negative defocusing was used to study whether decrease in beam density at the workpiece surface has a favorable effect on the formation of the weld bead. It can be seen from the images presented in Fig. 1.12 that decreasing the focal point position leads to a slight decrease in penetration. FPP − 4 mm resulted in full penetration in the set-up with the 600 μm process fiber, while none of the welds produced with the 300 μm fiber had complete penetration at the weld root [34].

The adoption of laser-welded profiles will have significant impacts on ship design, construction, and service life in the following aspects. The freedom from restrictions of profile types allows the optimization design of structural components based on

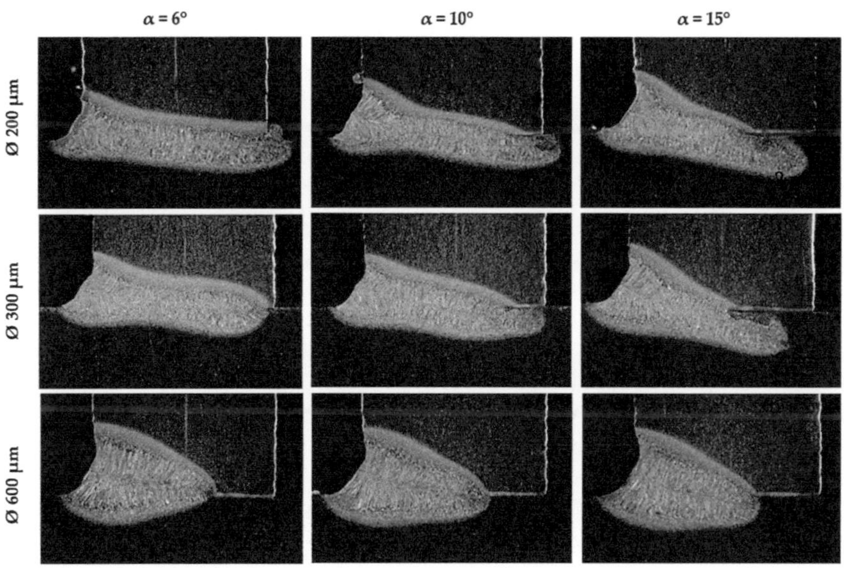

Fig. 1.11 Macrographs of weld samples at different beam inclination angles: AH36, $t = 8$ mm, $P_L = 6$ kW, $V_w = 1.25$ m/min, beam offset from flange 0.5 mm, FPP = -2 mm [34]

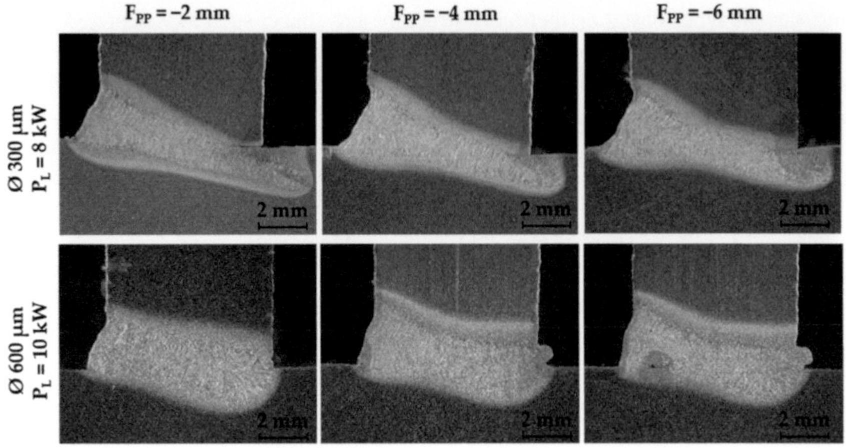

Fig. 1.12 Weld profiles at various focal point positions: AH36, t = 8 mm, V_W = 1.25 m/min, beam offset from flange 1.0 mm (up) and 1.5 mm (down), $\alpha = 15°$ [34]

performance as the primary consideration. Builders can transform complex three-dimensional components into simpler two-dimensional components, using high-power lasers to weld them faster and more accurately as a whole, thereby reducing construction costs, reducing finishing workloads, and reducing weight. For users, the use of welded profiles can significantly reduce operational costs. The fatigue life of components is extended due to improved processing accuracy and reduced welding seams, making it easier to rearrange the internal spaces of ships based on the needs of weapon equipment, and reducing fuel consumption, etc.

1.4.2 Application of Laser Welding in Aerospace

(1) The application of laser welding in aircraft manufacturing

In the field of aircraft manufacturing, laser welding technology is primarily used for the assembly of large skin panels and the welding of skin-to-stringer joints. Laser beam welding technology is also extensively used in the assembly of fuselage attachments, such as wing boxes for belly fins and flaps. In recent years, laser welding technology has also been widely used in the manufacturing of thin-walled components, such as intake ducts and corrugated pipes.

As early as the early 1970s, the United States used a 15 kW CO_2 laser for aircraft manufacturing in a variety of materials, components, welding tests and evaluation process standardization. Italy in the late 1970s from the United States introduced a 15 kW CO_2 laser, followed by the European Union on the aero-engine in a variety of containers and lightweight structure of the project, to carry out up to eight years of laser welding applications, materials involved in titanium alloys, nickel-based, iron-based high-temperature alloys, etc. German Aerospace Corporation (MBB) will be used for laser welding of aircraft fuselage, wings and internal bulkheads and reinforcement of all connections, replacing the original riveting process, the German aerospace industry known as a major technological revolution in the aviation manufacturing industry.

Around 2002, Airbus adopted the dual laser beam dual-sided synchronous laser welding process, to weld the skin and stringers of the lower sidewall panels of the A318 aluminum alloy aircraft fuselage into a unified fuselage panel. This resulted in a weight reduction of nearly 20% and increased strength by nearly 20% for the aircraft fuselage. This technology was subsequently widely applied in the manufacturing of fuselage panels for various models such as the A380, A340, and A350, as well as in the manufacturing of rib plates for cabin doors and unmanned aerial vehicle structures for the A380. In the case of the A350, the total length of weld seams reached 1000 m.

Laser welding technology research in China started in the 1980s. After years of effort, it has successfully achieved applications in automotive manufacturing, such as body panel welding and pipeline welding. With the increasing depth and breadth of research on laser welding processes for aluminum alloy and titanium alloy, laser welding applications for certain aircraft components have been realized. Such as the turbine disks, compressor blades, airframe and space capsule structural components, rings for jet engines, pressure vessels, rocket engine cases, helicopter rotor hubs, fasteners, engine exhausts and support flanges in Fig. 1.13.

(2) **In the field of aircraft repair**

Laser welding technology, characterized by fast welding speed and simpler construction compared to riveting, is an ideal alternative to riveting processes and has significant advantages in the field of aircraft structure repair. Some research has been conducted domestically on the repair of aircraft components using laser welding technology, and the repair application of certain components has been realized. A domestic aircraft repair facility has applied laser welding technology to the repair of damaged titanium alloy nacelles, and has developed process specifications for laser welding repair of titanium alloy structures. Around 2012, the First Institute of Aeronautics of the Air Force used a robotic flexible laser cutting/welding system to

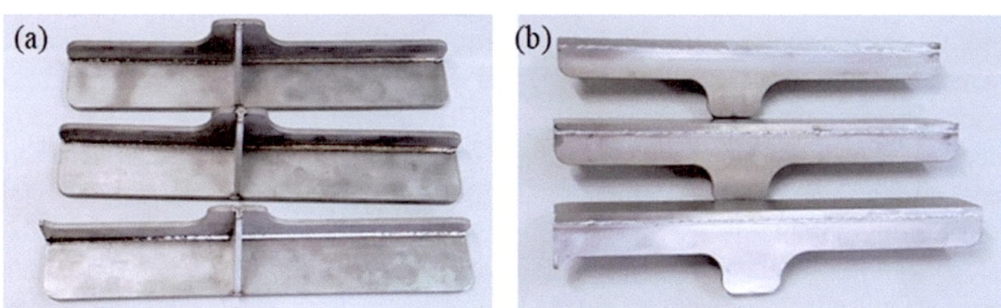

Fig. 1.13 Application drawing of laser welding aerospace support flanges [35]

complete the integrated research on cutting and welding repair of damaged areas on the underside of a certain aircraft's wing titanium alloy skin. The repaired surface was smooth, and the structural performance was fully restored, enabling long-term reliable operation and meeting usage requirements.

Additionally, laser welding has been used in the repair of engine components, such as turbine blades. The high precision and low heat input of laser welding make it suitable for repairing delicate and heat-sensitive parts, ensuring the performance and reliability of the repaired components. And laser welding technology has been applied in the repair and modification of aircraft skin panels, enabling smooth and seamless repairs without compromising the integrity and aerodynamic properties of the aircraft.

The application of laser welding in aircraft repair continues to expand and advance, contributing to the improvement of repair efficiency, cost-effectiveness, and the overall safety and performance of aircraft systems.

1.4.3 Application of Laser Welding in Automobile Manufacturing

From large work pieces such as automobile outer panels to minute parts such as wires and pins in electronic components, laser welding is used in a wide variety of automobile manufacturing.

(1) Spot welding on automobile frames/bodies

Laser-based spot welding is done by irradiating a laser beam on one side of a base material. Unlike resistance spot welding, there is no need to sandwich the base material between electrodes. This makes it possible to perform welding on a robot arm with a high degree of freedom for use even in intricate and complicated locations.

(2) Seam welding on automobile bodies

Using a high-speed pulse laser or continuous wave laser makes it possible to weld in lines rather than in spots. Line welding offers improved strength, and steel material for reinforcement is not required, allowing for reduced weight and higher rigidity. Laser brazing offers diversified advantages for sheet metal joining of automobile assemblies. The shows lower thermal distortion, high-quality surfaces with no additional sealing required, easy automation and simplified clamping/fixtures, enhanced design freedom, increased productivity etc. The laser-brazed joint considered across was located on a car roof and joined two 0.7 mm-thick different steel panels: the roof panel and the side panel, as depicted in Fig. 1.14a. The filler material was a silicon bronze alloy commonly used in these joints. Figure 1.14b illustrates the possible defects that may occur in such joints. Because of the difference between the panels' materials, the non-flat geometry and the limited accessible surface, the joint presents a very demanding inspection scenario [36].

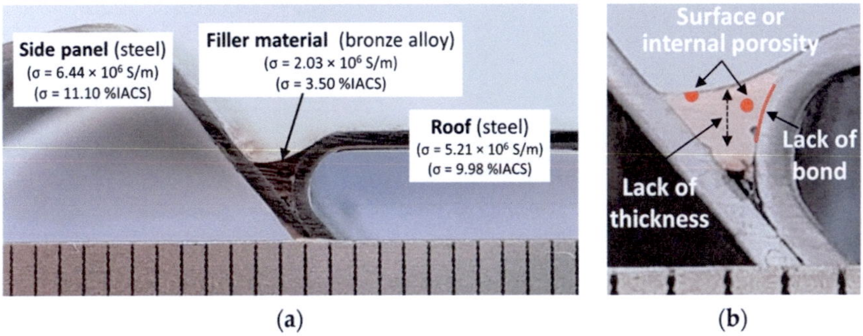

Fig. 1.14 **a** Laser-brazed weld profile section between the side panel and the roof steel sheets, **b** detail of possible defects morphology and location [36]

1.4.4 Application of Laser Welding in Microelectronic Components

Laser welding is a widely used technology in microelectronic component manufacturing due to its precision, speed, and versatility. Here are some applications of laser welding in microelectronic components:

(1) Package sealing: Laser welding is used to seal microelectronic packages, such as integrated circuits (ICs) and microprocessors. It provides a hermetic and reliable seal, ensuring protection against moisture, humidity, and other environmental factors that can damage the delicate electronics.

    Ultrafast lasers have been widely used in the fabrication of glass-based microdevices. Ultrafast laser micromachining provides advantages that are particularly useful for glass micromachining. Figure 1.15a shows the laser-welded sensor with good optical contact. Ultrafast laser welding can be accomplished by two approaches, continuous welding or point welding. The hermeticity of the sensor was the highest priority and thus we applied a continuous welding method to ensure a hermetically sealed sensor. Figure 1.15b shows that two welding seams around the inner components of the sensor were formed, and the interference fringes were selectively removed from the laser-welded regions, indicating that the bonding was successful and the interfacial gaps were completely removed [37].

(2) Wire bonding: Laser welding is used for wire bonding, which is the process of connecting wires to the bonding pads on semiconductor chips. The high precision and controlled heat input of laser welding make it ideal for bonding thin wires such as gold or aluminum wires, ensuring reliable electrical connections.

(3) Micro-sensors: Laser welding is employed in the fabrication of micro-sensors, such as pressure sensors, temperature sensors, and accelerometers. It allows precise assembly of sensor components and ensures a strong and stable bond between different materials, promoting sensor performance and reliability.

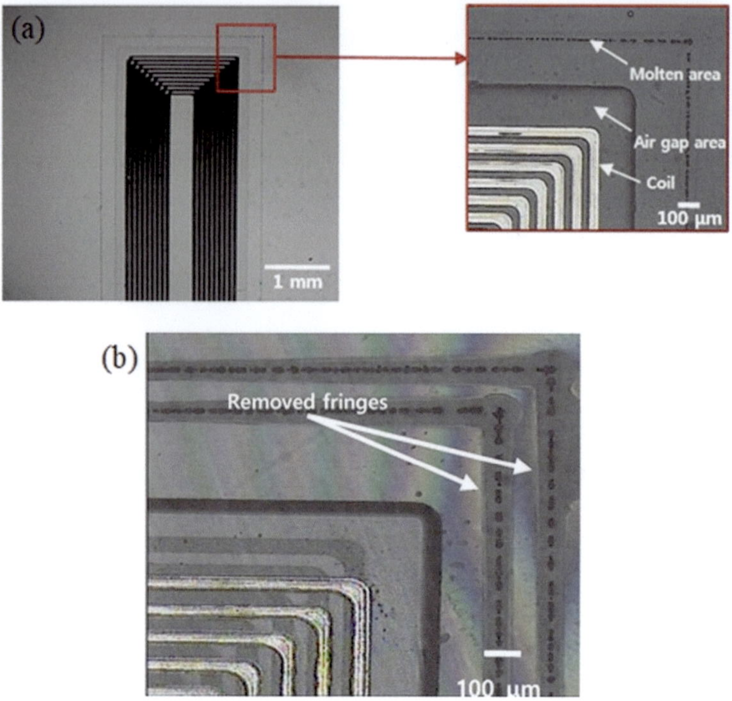

Fig. 1.15 **a** An optical image of the sealed implantable blood pressure sensor using ultrafast laser micro-welding, **b** zoomed image of the corner indicating successful welding [37]

(4) Optoelectronic devices: Laser welding is utilized in the assembly of optoelectronic devices, including laser diodes, photodiodes, and optical fibers. It enables precise alignment and bonding of optical components, ensuring efficient light coupling and reliable optical performance.
(5) Microelectronics packaging: Laser welding is employed in the assembly and packaging of microelectronic components, such as transistors, capacitors, and resistors, onto printed circuit boards (PCBs). It allows for high-density packaging, precise positioning, and efficient production processes.
(6) Micro-connectors: Laser welding is used for the assembly of micro-connectors, ensuring reliable electrical connections between different microelectronic devices and modules. It allows for high bonding strength, small joint size, and minimal heat affected zone, meeting the requirements of miniaturized electronic systems.

Overall, laser welding plays a crucial role in various aspects of microelectronic component manufacturing, offering precise, reliable, and efficient assembly solutions, thereby enhancing the performance and miniaturization of electronic devices.

1.5 Other Laser Processing Methods

1.5.1 Laser Cladding

Laser cladding is a hard facing process that uses a high-powered laser beam to melt the coating material and a thin layer of the substrate to form a low dilution, pore, and crack-free coating that is perfectly bonded to the substrate. The laser cladding experiment equipment is shown in Fig. 1.16. The process may be used for large area coverage by overlapping individual tracks, but it is the ability to protect smaller, localized areas that makes it unique. The wide range of materials that can be deposited and its suitability for treating small areas make laser cladding particularly appropriate to tailor surface properties to local service requirements and it opens up a new perspective for surface engineered materials. Laser cladding processes differ in the way the coating material is supplied. In two-step processes, a thin layer of material is deposited on the substrate, for example, as slurry or a thermally sprayed coating, and subsequently melted with the laser beam. In single-step processes, the material is fed continuously into the laser generated melt pool, usually in the form of a powder. Laser cladding by powder injection is superior to alternative processes and is the only one that has found practical use, because it is more energy efficient and it allows for better process control and reproducibility. Laser cladding has found relatively widespread use for the protection of materials against wear, corrosion and oxidation, for the deposition of self-lubricating coatings and thermal barriers, and for the refurbishing of high cost industrial components. Other applications with considerable potential are non-equilibrium synthesis of advanced materials, alloy development and free-form near-net shape manufacturing [38].

Laser cladding uses a high power laser as a heat source to form a cladding layer on the processing substrate. According to the powder feeding method, it can be classified into four types: coaxial powder system, preplaced powder system, off axis powder system and wire feeding system. The most commonly used LC methods are the coaxial powder system and the preplaced powder system. Figure 1.17 is a schematic diagram of the coaxial powder system and the preplaced powder system. The laser beam irradiates the substrate to form a liquid melt pool when the powder is ejected from the powder feeding nozzle by the carrier gas. After interacting with the laser, the powder enters the liquid melt pool and forms a cladding layer as the powder feeding nozzle moves synchronously with the laser beam. Different from the coaxial powder system, the cladding material is preplaced on the substrate in the preplaced powder system [40, 41]. Then, the preplaced powder is melted by scanning with a laser beam and the melt pool rapidly cools to form a cladding layer. LC samples can usually be divided into four parts: cladding zone (CZ), interfacial zone (IZ), heated affected zone (HAZ) and the substrate (SUB) [42]. In general, the preplaced powder system is simple on operation and has better cladding quality, but the penetration depth is not easy to control and the dilution is large. The coaxial powder system has a high laser utilization rate, but it has high requirements for the quality of cladding equipment.

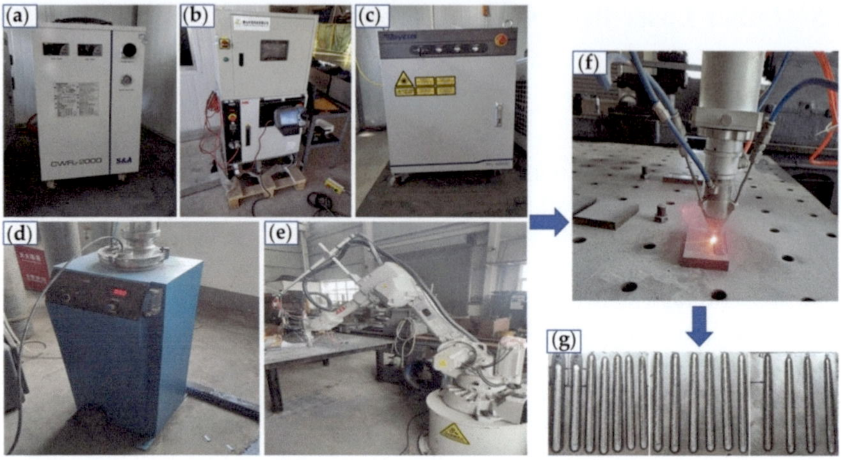

Fig. 1.16 Experimental equipment **a** cooling system, **b** control cabinet, **c** laser generator, **d** synchronous powder feeder, **e** ABB five axis manipulator, **f** cladding process, **g** laser cladding specimens [39]

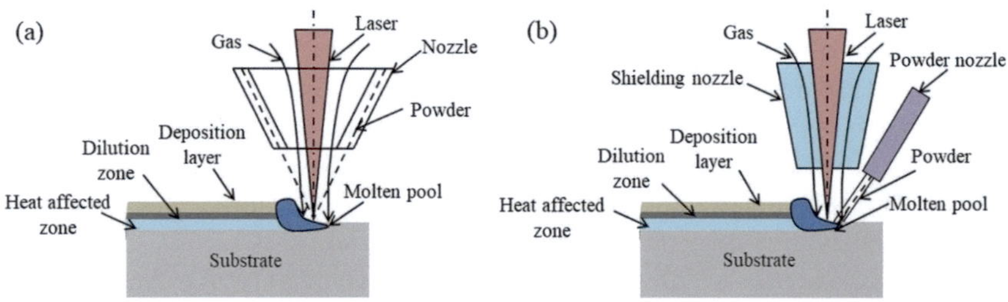

Fig. 1.17 Schematic diagram of different powder feeding systems: **a** coaxial powder feeding, **b** side-shaft powder feeding

Laser cladding is widely used in aerospace, chemical machinery and other fields because of its low dilution rate and good metallurgical bonding with the substrate. It can not only be used to form a surface-reinforced coating with good performance such as better wear resistance, better corrosion resistance and high temperature oxidation resistance, but also do good favor for us to repair machinery parts.

1.5.2 Laser Cutting

The schematic diagram of laser cutting is shown in Fig. 1.18. Laser cutting is a technology that utilizes a high-powered laser beam to precisely cut various materials. By focusing the laser beam onto the material's surface, it melts, burns, or vaporizes the

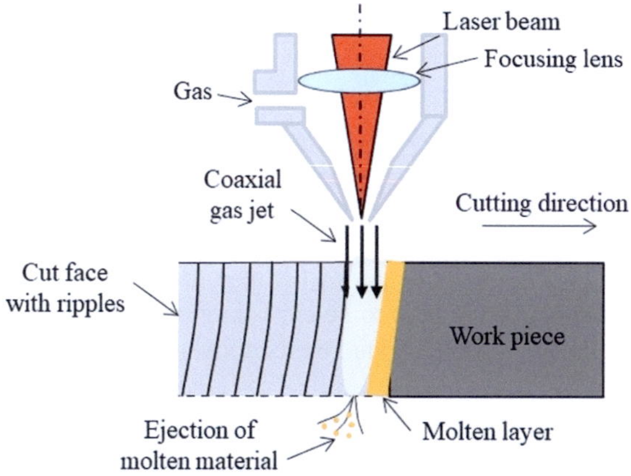

Fig. 1.18 Schematic diagram of laser cutting

material, enabling clean and intricate cuts. The advantages of laser cutting include its exceptional precision and accuracy, versatility for different materials, fast processing speed, minimal distortion, and the absence of tool wear. However, laser cutting also has some drawbacks, such as the high initial and operating costs, limitations with certain materials and thicknesses, safety concerns, and the need for skilled operators. Overall, laser cutting offers great potential for efficient and precise cutting operations.

Laser cutting is one of the most widely used thermal energy based non-contact type advance machining processes which can be applied to almost the whole range of metallic materials, like titanium, stainless steel, aluminum and alloys of aluminum, and furthermore to non-metallic materials such as wood, glass, plastics, ceramics, and composites utilized in various fabricating industries. The width of laser cutting, quality of the cut edges and the operating cost are affected by laser power, cutting speed, assist gas pressure, nozzle diameter and focus point position as well as the work-piece material.

1.5.3 Laser Drilling

Laser drilling is a non-contact process which utilizes the laser beam's capability of being focused on an intense spot to either melt or vaporize the material and thus forms a hole. Generally, there are four different approaches to laser drilling, namely single pulse, percussion, trepanning and helical drilling.

(1) Single-pulse/shot laser drilling

As the name suggests, in single-pulse laser drilling, a single shot of laser beam is fired at the material to create the required hole. In this method, both the source of the laser and the work material are kept static.

(2) Percussion laser drilling

In percussion drilling, a series of laser pulses are shot at the work material repeatedly. During this process, there is no relative motion between the laser beam and the work piece. This form of laser drilling effectively creates deeper and more precise holes with smaller diameters as compared to single-shot laser drilling.

(3) Trepanning

Trepanning is the process in which a laser beam is guided around a predefined locus. This locus is the center of the hole to be drilled. It is used in situations where the diameter of the hole to be cut is larger than the laser beam diameter. The accuracy of the hole is determined by the movement of the beam.

(4) Helical laser drilling

Helical laser drilling is a method in which the laser beam follows a helical path for movement while rotating about its own axis with respect to the work piece. A dove prism controls the movement of the laser beam.

Laser drilling process has become an industrial solution for massive micro or small hole drilling, high aspect ratios (up to 1:200) and a wide range of work piece materials. The main advantage of this process is the high productivity in comparison with mechanical drilling, being able to reach hundreds of holes per second in certain conditions. This operation is usually performed by applying a single pulse or a sequence of relatively short pulses (typically from 0.01 to 100 ms) and peak powers up to 50 kW. However, laser pulses up to 100 J can be achieved and pulses from femtoseconds to milliseconds can be applied [43, 44]. The combination of the pulse peak power and duration significantly influences the material removal mechanism. For ultra-short pulses (less than 100 fs) and high energy density (more than 10^{16} W/cm^2), material is removed by the cold ablation mechanism, since the material interacts with the energy of the laser and there is virtually no melted material. On the other hand, for longer pulses (more than 100 μs) and lower energy density (less than 10^7 W/cm^2), the material is removed by the melt expulsion mechanism, combining material vaporization and melting. The femtosecond laser machining experimental setup and the drilling method are shown in Fig. 1.19.

1.5.4 Laser Additive Manufacturing

Laser additive manufacturing (LAM), a novel manufacturing technique of layer-by-layer forming according to a geometric model, provides a decent option for materials processing. It owns the advantages of rapid prototyping, customization, high material utilization, and the ability to form complicated structures [46]. LAM contains two types: synchronous powder feeding (or wire feeding) forming method represented by laser directed energy deposition (LDED) and powder bed forming method represented by laser powder bed fusion (LPBF). The forming of high performance parts can be realized by LDED technology whose mechanical properties are equivalent to

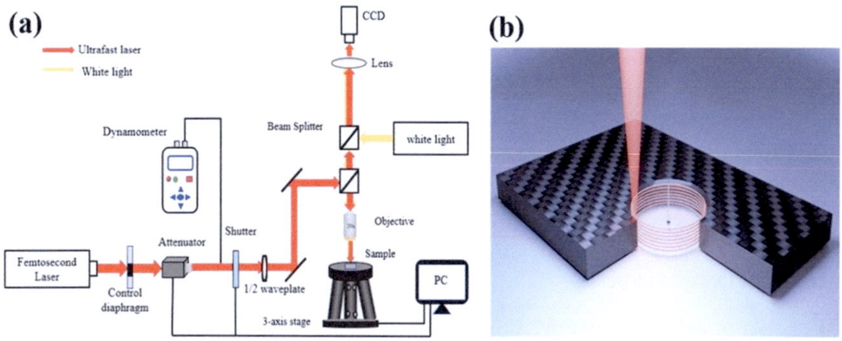

Fig. 1.19 **a** Schematic diagram showing the experimental setup, **b** schematic diagram showing laser drilling [45]

forgings. Besides LDED method and LPBF method can be used for manufacturing precision parts of complex shape.

(1) Laser directed energy deposition

The schematic of LDED system is shown in Fig. 1.20. LDED is an additive manufacturing process in which a laser beam forms a melt pool on a metallic substrate, into which powder is fed. The powder melts to form a deposit that is fusion-bonded to the substrate. The required geometry is built up in this way, layer by layer, the 3D part is created. Both the laser and nozzle from which the powder is delivered are manipulated using the gantry system or the robotic arm. The powder delivered into the same spot is absorbed into the melt pool, thus generating a deposit which may range from 0.1 to 0.5 cm in thick the resulting deposits may be used to build or repair the metal parts for many different applications.

LDED technology has a higher deposition rate and shortens the production cycle availably. What's more, LDED can be utilized to make parts with complex surfaces instead of some conventional machining technology, which can solve a series of problems machining hard, material wasting, tool wear etc. The ability to realize the preparation of parts with hollow structures and material gradients is undoubtedly one

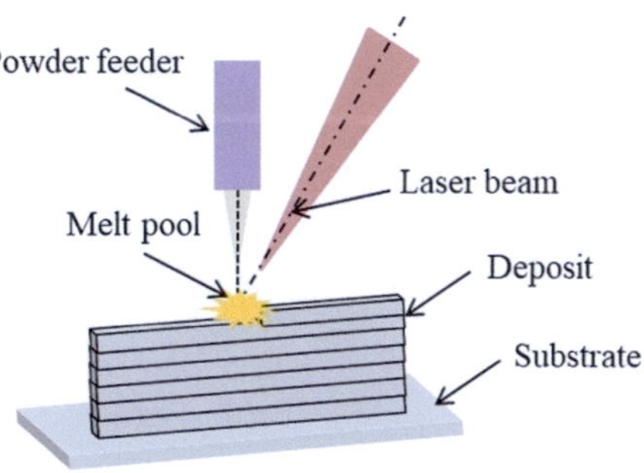

Fig. 1.20 Schematic diagram of the laser directed energy deposition (LDED) system

of the significant advantages of LDED. The components fabricated by LDED generally have compact texture and excellent material properties without heat-treatment. However, material utilization of LDED is low (about 1/4 to 1/3) while the use ratio of SLM is nearly 100%. Meanwhile, thermal stress is high in LDED processing resulting in interlinear cracking inside the work piece. The average particle size, spot diameter and layer thickness may result in worse size accuracy and higher surface roughness of LDED components than SLM. Hence, LDED combined with subsequent machining can improve manufacturing accuracy of parts, which makes parts meet work condition needs.

(2) Laser powder bed fusion

The schematic diagram of the LPBF technique is shown in Fig. 1.21. Laser powder bed fusion (LPBF), also referred to as Selective laser melting (SLM) or direct metal laser melting (DMLM), is an AM technique developed to melt and fuse metallic powders via a high power-density laser. The principle of the LPBF process starts with a building platform applied with very thin layers of metallic powders, which are completely melted later by the thermal energy induced by one or several laser beams. The cross-section area of the designed 3D part is built by selectively melting and re-solidifying metallic powders in each layer. The building platform is then lowered by a small distance and a new layer of powders is deposited and leveled by a re-coater. The laser beam can be directed and focused through a computer-generated pattern by carefully designed scanner optics. Therefore, the powder particles can be selectively melted in the powder bed and form the shape of 3D objects according to the CAD design [47].

The LPBF process generally uses powders with particle sizes between 20 and 50 μm and with a layer thickness of 20–100 μm. A large variety of metal powders have been revealed in the LPBF technique, mainly Zn, Mg, Cu, Cr, Co–Cr, stainless steel, Ni, Al, and Ti alloy. LPBF provides many advantages over conventional processes, such as the ability to create complex geometries with internal cavities or features without specific dies or tools, reduced time from design to testing, reduced need for assemblies and joining processes. Nevertheless, there are some problems

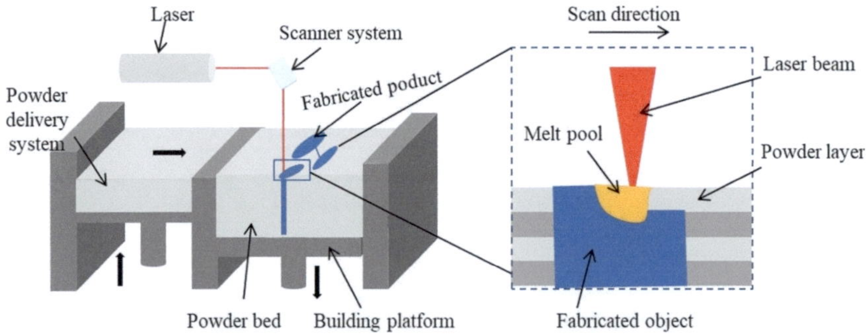

Fig. 1.21 Schematic diagram of the LPBF process

encountered in LPBF process. The major problems are high surface roughness and low part accuracy affected by stair-effect. Moreover, the high thermal gradients reinforce layered residual stresses due to full melting and solidification in a very short time. Because a large degree of shrinkage tends to occur during liquid-solid transformation, accumulating considerable stresses in LPBF process. The residual stresses arising during cooling are regarded as key factors responsible for the distortion and even delamination of the final products. Besides that, the instabilities of molten pool may result in periodization of the liquid molten pool and attendant interior porosity.

In general, LPBF end products tend to be stronger as they have fewer or no voids. A common use for LPBF printing is with 3D parts that have complex structures, geometries and thin walls. The aerospace industry uses LPBF 3D printing in some of its pioneering projects. These are typically those which focus on precise, durable, lightweight parts. LPBF is quite widespread now among the aerospace and medical orthopedics industries.

1.6 Summary and Prospects

Laser welding is a high-energy density welding technique characterized by its efficiency, precision, non-contact nature, and small heat-affected zone. It offers unique advantages for welding materials such as large thick plates, titanium-aluminum alloys, and dissimilar metals. Firstly, laser welding provides high-quality welds with no porosity or inclusions, ensuring weld integrity. Secondly, it is suitable for high-melting-point materials like titanium-aluminum alloys, allowing for high-strength connections. Additionally, laser welding excels in joining dissimilar metals, enabling the creation of multi-material composite structures. Laser welding has been successfully applied in various fields including automotive manufacturing, aerospace, electronics, medical devices, and metal fabrication. In automotive manufacturing, laser welding is widely used for body welding and engine component welding, improving welding speed and quality. In the aerospace industry, it ensures strength and sealing through welding applications in aircraft engines and spacecraft structures. In electronics, laser welding is utilized for micro-welding tasks such as electronic components, batteries, and circuit boards. In medical device manufacturing, laser welding is extensively employed for welding titanium implants and medical instruments. Furthermore, laser welding plays a significant role in the fabrication of pressure vessels, pipelines, and structural components in the metal industry. In summary, the efficiency, precision, and versatility of laser welding have facilitated successful applications in multiple industries, driving advancements in industrial manufacturing.

(1) Automation and intelligence: With the rapid development of automation technology and artificial intelligence, laser welding will be more integrated with robotic technology to achieve automated welding processes. This will improve

production efficiency, reduce labor intensity for workers, and ensure welding quality and consistency.

(2) Welding of new materials: With the emergence of new materials such as high-strength steel, composite materials, and new alloys, laser welding will face more application challenges and opportunities. Laser welding has unique energy control and precision, making it suitable for welding various new materials and driving the development of new material applications.

(3) Micro welding and micro-processing: With the rapid development of micro-processing fields such as microelectronic devices and optoelectronic devices, the demand for micro welding technology is also increasing. The high energy density and precise control capabilities of laser welding make it an ideal choice for micro welding, enabling reliable connections of small-sized components.

(4) Applications in new fields: Laser welding has been successfully applied in multiple fields such as automotive, aerospace, electronics, medical, and metal manufacturing. With further technological advancements, laser welding is expected to have a wider range of applications in other fields such as energy, construction, and shipbuilding.

In conclusion, laser welding has vast development prospects. With continuous technological innovations and expanded application domains, laser welding will continue to play an important role in industrial manufacturing, enhancing efficiency, quality, and innovation capabilities.

Problems

1. Briefly introduce the principles and characteristics of laser.
2. Briefly describe the advantages and disadvantages of laser welding.
3. Briefly introduce the classification of laser welding equipment and their corresponding application fields.
4. Please explain the principle and difference between deep penetration welding and heat conduction welding in laser welding.
5. In laser welding, welding speed has a significant impact on welding quality. Please explain why increasing welding speed may lead to a decrease in weld seam quality.
6. What type of laser is typically used in laser welding? Please describe its characteristics and explain why it is suitable for laser welding.
7. Laser welding has many advantages, but it also has some limitations. Please list at least three limitations of laser welding and discuss their impact on applications.
8. When laser welding, the quality of the weld seam is influenced by many factors. Please list at least five factors that may affect weld seam quality and briefly explain their impact.
9. The differences between laser welding and traditional arc welding in terms of principles, characteristics, and applications.
10. In which industries is laser welding widely applied? Please provide at least three application areas and briefly explain their application situations.

References

1. Peças P, Henrique M, Miranda RM et al (1995) Laser welding of low-thickness zinc-coated and uncoated carbon steel sheets. Opt Quant Electron 27(12):1193–1201
2. Tambasco J, Anthony T, Sandven O (1996) Laser welding in the dental laboratory: an alternative to soldering. J Dent Technol Peer Rev Publ Natl Assoc Dent Lab 13(4):23–31
3. Bennett WR (2000) Background of an inversion: the first gas laser. IEEE J Sel Top Quantum Electron 6(6):869–875
4. Jain S, Vibhute PK, Patil C et al (2020) Laser welding in orthodontics: a review study. J Dent Health Oral Res 1(1):1–14
5. Kovacs T (2018) Laser welding process specification base on welding theories. Procedia Manuf 22:147–153
6. Venkat S (2009) Welding with high power diode lasers. Laser Tech J 6(2):31–32
7. Buddu RK, Chauhan N, Raole PM et al (2015) Studies on mechanical properties, microstructure and fracture morphology details of laser beam welded thick SS304L plates for fusion reactor applications. Fusion Eng Des 95:34–43
8. Zhang MJ, Chen GY, Zhou Y et al (2014) Optimization of deep penetration laser welding of thick stainless steel with a 10kW fiber laser. Mater Des 53:568–576
9. Li YT, Geng SN, Shu LS et al (2023) Ultra-high-power laser welding of thick-section steel: current research progress and future perspectives. Opt Laser Technol 167:109663
10. Sharma SK, Maheshwari S (2017) A review on welding of high strength oil and gas pipeline steels. J Nat Gas Sci Eng 38:203–217
11. Li JZ, Sun QJ, Kang KX et al (2020) Process stability and parameters optimization of narrow-gap laser vertical welding with hot wire for thick stainless steel in nuclear power plant. Opt Laser Technol 123:105921
12. Wu RL, Huang Y, Rong YM et al (2023) Solid phase transformation effects on stress and strain in the thick plate EH40 welded butt joint. Opt Laser Technol 157:108635
13. Balakrishnan J, Vasileiou AN, Francis JA et al (2018) Residual stress distributions in arc, laser and electron-beam welds in 30 mm thick SA508 steel: a cross-process comparison. Int J Press Vessels Pip 162:59–70
14. Kong HJ, Xu C, Bu C et al (2019) Hardening mechanisms and impact toughening of a high-strength steel containing low Ni and Cu additions. Acta Mater 172:150–160
15. Kawahito Y, Wang HZ, Katayama S et al (2018) Ultra high power (100 kW) fiber laser welding of steel. Opt Lett 43(19):4667–4670
16. Katayama S, Kawahito Y, Mizutani M (2010) Elucidation of laser welding phenomena and factors affecting weld penetration and welding defects. Phys Procedia 5:9–17
17. Trowell KA, Goroshin S, Frost DL et al (2020) Aluminum and its role as a recyclable, sustainable carrier of renewable energy. Appl Energy 275:115112
18. Hong KM, Shin YC (2017) Prospects of laser welding technology in the automotive industry: a review. J Mater Process Technol 245:46–69
19. Zwicker MFR, Moghadam M, Zhang W et al (2020) Automotive battery pack manufacturing–a review of battery to tab joining. J Adv Join Process 1:100017
20. George EP, Raabe D, Ritchie RO (2019) High-entropy alloys. Nat Rev Mater 4(8):515–534
21. Wang JM, Peng GC, Li LQ et al (2019) 30 kW-level laser welding characteristics of 5A06 aluminum alloy thick plate under subatmospheric pressure. Opt Laser Technol 119:105668
22. Bhat BN, Pandey AB, Tamirisakandala S et al (2018) Aerospace materials characteristics: aerospace materials and applications. American Institute of Aeronautics and Astronautics, New York
23. Quazi MM, Ishak M, Fazal MA et al (2020) Current research and development status of dissimilar materials laser welding of titanium and its alloys. Opt Laser Technol 126:106090
24. Sun Z, Ion JC (1995) Laser welding of dissimilar metal combinations. J Mater Sci 30(17):4205–4214
25. Zhang PL, Shi HC, Tian YT et al (2021) Effect of zinc on the fracture behavior of galvanized steel/6061 aluminum alloy by laser brazing. Weld World 65(1):13–22

26. Fang YJ, Jiang XS, Mo DF et al (2019) A review on dissimilar metals' welding methods and mechanisms with interlayer. Int J Adv Manuf Technol 102(9):2845–2863
27. Geng YC, Akbari M, Karimipour A et al (2019) Effects of the laser parameters on the mechanical properties and microstructure of weld joint in dissimilar pulsed laser welding of AISI 304 and AISI 420. Infrared Phys Technol 103:103081
28. Sharifi M, Akbari M (2019) Experimental investigation of the effect of process parameters on cutting region temperature and cutting edge quality in laser cutting of AL6061T6 alloy. Optik 184:457–463
29. Miller WS, Zhuang L, Bottema J et al (2000) Recent development in aluminium alloys for the automotive industry. Mater Sci Eng A 280(1):37–49
30. Dong HG, Hu WJ, Duan YP et al (2012) Dissimilar metal joining of aluminum alloy to galvanized steel with Al–Si, Al–Cu, Al–Si–Cu and Zn–Al filler wires. J Mater Process Technol 212(2):458–464
31. Yu GY, Chen SH, Zou TP et al (2020) Laser beam joining of Al/steel dissimilar metals with Sn-Zn filler wire in overlap configuration. J Manuf Process 60:481–493
32. Dimatteo V, Ascari A, Fortunato A (2019) Continuous laser welding with spatial beam oscillation of dissimilar thin sheet materials (Al-Cu and Cu-Al): process optimization and characterization. J Manuf Process 44:158–165
33. Yan SH, Shi Y (2019) Influence of laser power on microstructure and mechanical property of laser-welded Al/Cu dissimilar lap joints. J Manuf Process 45:312–321
34. Unt A, Poutiainen I, Grünenwald S et al (2017) High power fiber laser welding of single sided T-joint on shipbuilding steel with different processing setups. Appl Sci 7(12):1276
35. Caiazzo F, Alfieri V, Corrado G et al (2017) Laser beam welding of a Ti–6Al–4V support flange for buy-to-fly reduction. Metals 7(5):183
36. Machado MA, Rosado LS, Mendes NM et al (2021) Multisensor inspection of laser-brazed joints in the automotive industry. Sensors 21(21):7335
37. Kim S, Park J, So S et al (2019) Characteristics of an implantable blood pressure sensor packaged by ultrafast laser microwelding. Sensors 19(8):1801
38. Vilar R (1999) Laser cladding. J Laser Appl 11(2):64–79
39. Li TK, Long HY, Qiu CM et al (2022) Multi-objective optimization of process parameters of 45 steel laser cladding Ni60PTA alloy powder. Coatings 12(7):939
40. Wirth F, Wegener K (2018) A physical modeling and predictive simulation of the laser cladding process. Addit Manuf 22:307–319
41. Lei JB, Shi C, Zhou SF et al (2018) Enhanced corrosion and wear resistance properties of carbon fiber reinforced Ni-based composite coating by laser cladding. Surf Coat Technol 334:274–285
42. Liu K, Li YJ, Wang J et al (2015) Effect of high dilution on the *in situ* synthesis of Ni–Zr/Zr–Si(B, C) reinforced composite coating on zirconium alloy substrate by laser cladding. Mater Des 87:66–74
43. Schulz W, Eppelt U, Poprawe R (2013) Review on laser drilling I. Fundamentals, modeling, and simulation. J Laser Appl 25(1):102–137
44. Ng GKL, Li L (2001) The effect of laser peak power and pulse width on the hole geometry repeatability in laser percussion drilling. Opt Laser Technol 33(6):393–402
45. Li YY, He GY, Liu HL et al (2023) Investigation of heat accumulation in femtosecond laser drilling of carbon fiber-reinforced polymer. Micromachines 14(5):913
46. Gong GH, Ye JJ, Chi YM et al (2021) Research status of laser additive manufacturing for metal: a review. J Market Res 15:855–884
47. Li ZY, Huang MF, Zhong YR et al (2022) A description logic based ontology for knowledge representation in process planning for laser powder bed fusion. Appl Sci 12(9):4612

Chapter 2
Laser-Arc Hybrid Welding

2.1 Principle, Characteristics and Classification of Laser-Arc Hybrid Welding

The hybrid laser arc welding process was introduced in the late 1970s by Prof. W. M. Steen and co-researchers, and the process was termed as "arc-augmented laser welding". The results of their research showed the clear advantages of combining a laser beam and an electric arc for welding. The hybridization effect shows a remarkable increase in welding speed, penetration depth and process stability. After the successful demonstration of this technique, further research and development on laser-arc hybrid welding experienced a slow growth due to the lack of availability of high power, reliable laser source, required human skills, and incomplete knowledge of the process. The development of reliable high power industrial lasers in late 1980s [1], caught the attention of the researchers from industry and academics, to engage in the improvement of the process, to make it viable for industrial applications. Several drawbacks of the individual welding processes are eliminated, including the problems like gap bridging ability and reflectivity of the materials are no longer an issue for performing successful welding [2]. In the 1990s, the hybrid laser arc welding appeared greater developments due to the availability of high-power CO_2 laser. A number of studies on hybrid laser arc welding processes were carried out focusing on the economic viability of the process, welding position and preparation of edges, minimization of metallurgical defects, optimization of process parameters, etc.

However, after reporting of several process advantages, there are very limited industrial applications due to the high equipment cost and complexity of the process parameters, and the unavailability of an integrated commercial laser-arc hybrid welding system. The first industrial laser-arc hybrid welding system was introduced in 2000 by Fraunhofer ILT, Germany, in an oil tank manufacturing industry. This system has been installed in different industries including automotive industry, ship building and tube manufacturing industries for welding of metallic sheets [3]. A number of laser-arc hybrid welding systems are commercially available now. The

R. Li and T. Li, *Advanced Welding Methods and Equipment*,
https://doi.org/10.1007/978-981-97-4109-0_2

industrial applications of laser-arc hybrid welding have grown in this century due to not only the development of commercially available integrated hybrid welding heads, but also due to the availability of cheaper, reliable and efficient high power laser systems with the advancement of instrumentation, automation and control.

Global trends in manufacturing of welded components are characterized by a growing demand for reduced welding operation costs and welding cycle times on the one hand, and consistent, or even improved, quality and mechanical integrity of welded components on the other. In this situation, the substitution of a new welding technology for the existing traditional ones with a view to increasing productivity could be beneficial, if it guarantees the required quality levels. Laser hybrid welding technology provides a number of advantages, which over the last decade have seen it applied across a broad range of industrial branches such as automotive manufacturing, shipbuilding and power generation.

Hybrid welding, using the combination of a laser and an electrical arc, is designed to overcome problems commonly encountered during either laser or arc welding such as cracking, brittle phase formation and porosity. When placed in close contact with each other, the two heat sources interact in such a way as to produce a single high intensity energy source. This synergistic interaction of the two heat sources has been shown to alleviate problems commonly encountered in each individual welding process. Hybrid welding allows increased gap tolerances, as compared to laser welding, while retaining the high weld speed and penetration necessary for the efficient welding of thicker work pieces. A number of simultaneously occurring physical processes have been identified as contributing to these unique properties obtained during hybrid welding. However, the physical understanding of these interactions is still evolving.

2.1.1 Principle of Laser-Arc Hybrid Welding

Laser-arc hybrid welding is a welding technique that combines the precision of laser welding with the high heat input of arc welding. This innovative approach has gained attention in various industrial applications due to its ability to deliver exceptional weld quality and productivity. Below, we outline the key advantages and disadvantages of laser-arc hybrid welding:

There are some advantages:

(1) High Weld Quality: Laser-arc hybrid welding delivers excellent welding quality, including exceptional weld bead appearance, minimal porosity, low residual stresses, and high-density weld seams.
(2) High Welding Efficiency: This welding method combines the precision of laser welding with the high heat input of arc welding, allowing for increased welding productivity within the same timeframe.

(3) Versatility with Multiple Materials: Laser-arc hybrid welding is applicable to a wide range of materials, including high-alloy steels, stainless steel, aluminum alloys, and more, making it versatile across various industrial sectors.
(4) Low Heat-Affected Zone: Due to the high energy density of laser welding, this method can minimize the heat-affected zone, reducing the risk of material annealing and embrittlement.
(5) Precise Welding Control: Advanced automation and control systems enable highly precise control of the welding process, ensuring consistent welding quality.

Of course, there will be some disadvantages:

(1) High Initial Investment: Laser-arc hybrid welding requires expensive equipment and specialized technology, necessitating a significant initial investment.
(2) Complexity: Compared to traditional welding methods, this hybrid welding technology demands more complex equipment and operations, requiring trained operators.
(3) Environmental Sensitivity: Laser welding processes may produce harmful gases and optical radiation, necessitating appropriate safety measures to protect operators and the environment.
(4) High Joint Preparation Requirements: Elevated joint preparation requirements can increase production time and costs.
(5) Limitation on Material Thickness: Laser welding is typically applied to thinner materials, making it less suitable for thicker materials.

In summary, laser-arc hybrid welding is an advanced welding technology suitable for applications demanding high-quality welds. However, it's important to consider factors such as the high initial investment and complexity when deciding whether to adopt this technology. The choice should be based on a careful assessment of its advantages and disadvantages, in alignment with the specific requirements of the application.

2.1.2 *Characteristics of Laser-Arc Hybrid Welding*

In order to expand the range of applications for hybrid welding and to optimize the process for its current applications, the mechanisms governing the physical processes in hybrid welding must be better understood. There are several factors originating from both processes and combined effects. They have an effect on plasma characteristics since the heat distribution inside plasma differs from those of both sources separately, thus resulting in a synergetic effect on the process. Another effect is the preheating mechanisms, which depend on the order of the processes and their separation from each other. All of these features affect the welding process and properties such as the shape of the weld produced. Different research groups have also taken into use different set-ups for the process. A brief critical assessment is presented here to increase understanding of the hybrid welding process.

(1) **Plasma characteristics**

The interaction of the laser and the arc is a highly complex physical process in which metal vapor generated by the laser can greatly affect the electrical and thermal conductivity of the arc plasma and the behavior of the arc column. The interaction of the two heat sources significantly affects the resulting weldment properties.

The plasma phase absorbs a portion of the laser energy. The plasma phase is also formed by the interaction of the arc and the surrounding atmosphere. The size of the arc plasma often depends on the distance between the heat sources, and the current density, arc length and voltage. It has also been observed that the laser and arc plasmas interact, and the transmission of the laser through the plasma results in additional laser attenuation.

During hybrid welding, the arc undergoes contraction, in which its width is nearly the same size as the beam of the laser. This has been reported to be due to the presence of the laser-induced plasma, which causes the arc resistance and radius to decrease. The influence of laser-induced plasma on arc resistivity and stability can be explained by two phenomena. The metal vapor enhances current conduction and reduces arc voltage. Since the arc follows the path of least electrical resistance, the arc tends to bend, and the arc root forms within close proximity to the keyhole. It has been observed that the arc voltage decreases after switching on the laser beam. The reason might lie in the metal vapor produced when the laser beam is initiated.

The interaction between the laser beam and arc plasma has been studied, and the investigation results present three key features: arc plasma preheats the base metal and enhances the absorption of the laser beam by the metal, especially in the case of CO_2 laser; the arc plasma dilutes the laser-induced plasma and reduces the focal point position for the laser beam, and the laser beam stabilizes arc plasma. Since high-power laser beams are used, the physical characteristics of laser welding are hardly changed, so it is simply considered that the preheating effect on the base metal is the main effect of the arc plasma on the laser beam material interaction.

When the process distance is significantly longer than the arc plasma radius, the laser and arc plasmas are separate. In contrast, when the process distance between them is less than the arc plasma radius or approximately the same, two plasmas interact. Therefore, the interaction of the laser beam and the arc plasma strongly depends on the distance between the two heat sources. Furthermore, when the arc and laser are in close proximity, calculations show that the arc radius must be decreased to achieve the observed hybrid weld depth. The arc radius increases significantly beyond the process distance due to the lack of metal vapor entering the arc plasma, and the energy density of the arc decreases. This is due to the fact that the processes are now acting as two separate processes, i.e., as a tandem process. Hence, the cooling rate decreases with increasing process distance. Tandem welding occurs when the distance between the heat sources is significantly longer than the arc plasma radius, usually a process distance of 5–8 mm. In tandem welding, the arc and laser act separately on the work piece. A quantitative understanding of the process distance and its relationship with welding parameters will lead to a significant improvement in process control.

(2) **Synergistic effect of the welding arc and laser beam**

When the action of the welding arc and the laser beam are combined, the temperature of the anode spot and cathode spot increases greatly, leading to greater ionization. This is particularly noticeable in welding with electrode positive since the cathode spot is located on the work piece and its relative motion compared to the joint is less intensive and limited to a much smaller arc than in classical arc welding, for example, gas tungsten arc welding (GTAW) and plasma arc welding (PAW). A second very important feature is arc contraction. It is generally accepted that the greater the temperature difference between the arc center and the environment, the stronger the welding arc contraction. The arc welding process performance is improved with small process distance which greatly increases the temperature in the arc center while the temperature of the environment does not change. Similarly, the interaction of the arc and laser often leads to improved laser welding characteristics. Some metals strongly reflect laser light, which renders initiation of melting difficult. The welding arc, however, heats up the surface at the welding spot and permits the laser beam to transfer all its energy to the work piece. The cessation of high energy beam action results in solidification of the molten pool in the weld as a result of cohesive and other forces.

The synergistic action of the laser beam and welding arc shows that when arc current intensities are low, the laser beam affects ionization, reduces arc resistance, and increases the number of carriers of the electric current. This influence is understandably reduced when using higher current intensities. It is, however, not known which property of the laser beam contributes the most to the higher ionization, whether it is the higher concentration of heat energy alone or to some extent the presence of electromagnetic waves with short wavelength. The difference in the influences exerted by CO_2 and Nd: YAG lasers on the degree of ionization in the welding arc is also not evident from the literature.

The synergistic effects of the heat source combination are maximized when the laser beam is located between the arc center and droplet detachment of the electrode. It is understood that the final bead shape in hybrid welding is influenced by the volume of the molten pool before the impingement of the laser beam. This means that the critical depth of the molten pool must form before the irradiation of the laser beam to maximize the coupling effect of the heat sources observed as the distance between the laser and arc increases. The molten metal from the arc is not fully mixed as in laser welding. The optimum process distance in different process variations can be summarized as 1–5 mm. This suggests that the electric pole of the arc moves up as the distance between the laser and the arc increases. Thus, if the process distance is short, the laser leads the electric pole of the arc into deeper parts of the groove when an air gap exists. If the process separation increases, the interaction between the arc and laser weakens, causing the pole to move upwards toward the top of the groove.

(3) **Preheating mechanism**

The arc can preheat the work piece surface before laser irradiation and form a shallow melting zone which reduces the reflectivity of the work piece. Consequently, the laser energy absorbed by the work piece is increased, especially in the case of CO_2 lasers. The interaction of the laser-induced plasma and arc plasma is brought about by firstly diluting the laser-induced plasma with the arc plasma. The laser induced plasma is very different from the arc plasma in terms of characteristics and it has a much higher temperature and particle density. When plasma encounter spatially, the particles of the laser-induced plasma enter the arc plasma. As a result, the particle density of the laser-induced plasma is diluted and reduced. This reduces the plasma effect of the laser and enhances the laser energy effectively absorbed by the work piece.

The larger the heat source interaction, the greater the change of the melting energy. Therefore, the interaction between the two heat sources can be discussed in terms of a quantitative analysis of the changes in melting energy by various welding parameters. The role of different mechanisms in heat source interaction requires further study [4]. Compared with the total melting energy of laser welding and gas metal arc welding. The melting energy can be increased by up to 23% using an optimal combination of parameters in laser-arc hybrid welding.

2.1.3 *Classification of Laser-Arc Hybrid Welding*

The application of a variety of laser hybrid welding processes is wide and ranges from thin sheet metal welding to heavy section welding with plate thicknesses of up to 20 mm. Depending on the particular welding task, the choice of the appropriate laser or arc processes could be crucial to the successful application of the technology in terms of the welding process requirements (e.g. minimal or maximal required heat input, high melting efficiency of the welding process, required penetration depth, special filler materials, necessary dilution in weld metal, etc.). Recent reports state that hybrid processes with different laser sources like CO_2 and solid-state lasers as well as different arc processes (GMAW, GTAW, PAW, SAW) can be used in combination with lasers.

(1) **Laser-GMAW hybrid welding**

The principal diagram of laser-GMAW hybrid welding is shown in Fig. 2.1. In hybrid laser-GMAW process, a CO_2 laser, a Nd: YAG laser or a fiber laser is used as the primary heat source and the electric arc produced by GMAW is used as the secondary heat source. In this process, the arc is generated between the consumable metal electrode and the workpiece. The electrode is continuously fed from a spool. The laser generates a keyhole at the weld pool which ensures a deep penetration weld, whereas, the molten filler material produced during GMAW helps to fill the root gap and permits welding of the parts with higher gaps and fit-up issues [5]. It is observed that the stability of GMAW arc improves when it combines with a laser beam. The use of filler materials improves the chemical composition of the weldment by adding alloying elements in the weld pool [6] and also improves the mechanical properties of the weldment. The greater volume of melt pool produced during GMAW makes this process more efficient for bridging the weld gap than GTAW and PAW. Shielding gases are used to protect the molten weld pool from atmospheric contamination. Inert gases like helium, argon or nitrogen or a mixture of those gases in different proportions are generally used during GMAW, whereas, active gases like CO_2, O_2 or the mixture of CO_2 with inert gases in different proportions are used during MAG (metal active gas) welding, as shielding gases [7]. A pulsed mode GMAW power supply is generally used in hybrid laser GMAW which enables better control of arc length, formation of metal droplet and its transfer mode, arc stability and spatter formation during welding. Due to the fast-cooling rate of hybrid laser arc welding process, the electrode wire composition used in GMAW may not be acceptable for hybrid laser GMAW welding process [8]. The weld pool formed during hybrid laser GMAW welding has greater proportions of filler material at the top cap part and the base material at the bottom root part. A small joint gap up to 1 mm enables the molten filler to reach to the root of the joint and thus improve the degree of homogeneity by improving dilution in the weld pool. Because of the use of multiple heat sources in hybrid laser GMAW, the welding speed can be further increased than laser bean welding and GMAW, without the formation of undercuts [9].

(2) **Laser-GTAW hybrid welding**

The principal diagram of laser-GTAW hybrid welding is shown in Fig. 2.2. In hybrid laser-GTAW process, the electric arc produced during GTAW, is used as a secondary heat source. Because of the use of non-consumable tungsten electrode, additional filler materials may be required during welding. The rate of metal deposition during hybrid laser GTAW is much lesser than that of hybrid laser GMAW, because only a fraction of heat of the arc is used to melt the filler wire during hybrid laser GTAW process. As the filler wire is separated from the electrical circuit of the hybrid laser GTAW set up, the welding current and the metal deposition rate during welding can be determined independently. However, the electrode contamination due to hot metal vapor, the erosion of tungsten electrode tip due to the use of high frequency arc ignition, and the high affinity of tungsten to oxygen are the

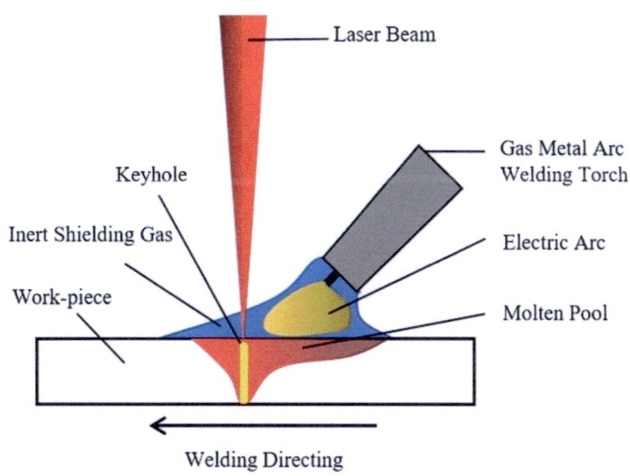

Fig. 2.1 Laser-GMAW hybrid welding principal diagram

common problems associated with GTAW process. Shielding gas is used to protect the weld pool and also to protect the electrode from atmospheric contamination or oxide formation. The melting rate of base material for hybrid laser GTAW is nearly twice the sum of the melting rate of laser welding and GTAW welding, individually. It is reported that when a GTAW arc is used with a laser beam in a common interaction zone, the laser energy absorption in the base material is improved due to the preheating of base material using leading GTAW arc. On the other hand, the use of laser at common weld pool stabilizes the GTAW arc[9]. Because of the high power density of laser beam, a part of base material at the weld zone melts and vaporizes, which provides a conductive path for the arc [10]. The GTAW arc follows the shortest conductive path between electrode and metal, thus roots to the laser interaction zone. The rooting and narrowing of the arc to laser interaction zone increase electrical conductivity along the arc column and stabilize the arc [11]. GTAW is generally used for welding hard-to-weld metals such as stainless steel, aluminum-magnesium-alloys. The hybrid laser-GTAW welding is found suitable for welding AISI304 stainless steel due to its high welding speed and excellent mechanical properties [12]. The hybrid laser-GTAW process is generally used for welding thin gauge materials. The ability of hybrid laser-GTAW for butt welding of very thin austenitic stainless steel sheets has been demonstrated [13].

(3) **Laser-PAW hybrid welding**

The principal diagram of laser-PAW hybrid welding is shown in Fig. 2.3. In the laser PAW hybrid welding process, a constricted plasma arc is used as the secondary heat source. Similar to GTAW welding, non consumable tungsten or tungsten alloy wire is used as the electrode in plasma arc welding. Filler materials may or may not be used. Unlike GTAW, the electrode used in PAW is recessed in the welding torch. A water-cooled constricting nozzle of small diameter is used inside the welding torch to squeeze the arc, which increases its intensity and improves the stability and shape of the arc. The inert gas passed through the constricting nozzle, known as orifice gas, is used to produce plasma. The shielding effect is further enhanced by using another

Fig. 2.2 Laser-GTAW hybrid welding principal diagram

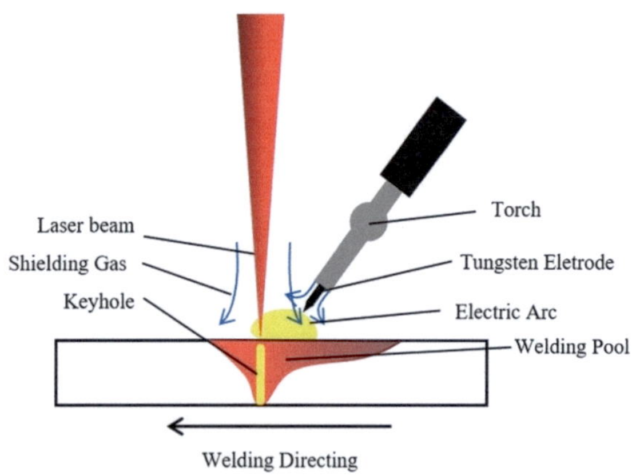

shielding gas, which is passed through the outer ring of the torch. In PAW, the arc is established either between the electrode and the constricting nozzle (nontransferable arc) or between the electrode and the workpiece material (transfer arc). Initially, high-frequency current is used during the ignition of the pilot arc between the electrode and the constricting nozzle, and the current is fixed at a level after the pilot arc is established. During welding, the arc is transferred to the workpiece, which turns into part of the electrical circuit. As compared to the GTAW arc, the plasma arc is more stable and pointed. The plasma arc does not deflect easily at the closest point of base metal because of the jet action due to the contraction constriction of the arc plasma. The laser beam and plasma jet are directed at the common interaction zone. The laser beam is either directed perpendicular to the material surface or it is tilted at a small angle from the perpendicular axis. The PAW torch is generally held at an angle of 35–45 degrees from the laser beam axis in a paraxial arrangement [14]. In a coaxial arrangement, the laser beam is surrounded by the plasma arc. The hybridization effect of laser and plasma arc increases the welding speed and stabilizes the arc plasma. The penetration depth is more for laser PAW hybrid welding than for laser welding for the same energy input per unit length of the weld [15]. As compared to the results of laser-GTAW hybrid welding, the heat input from the arc can be reduced up to 40% when laser PAW hybrid welding is used [16].

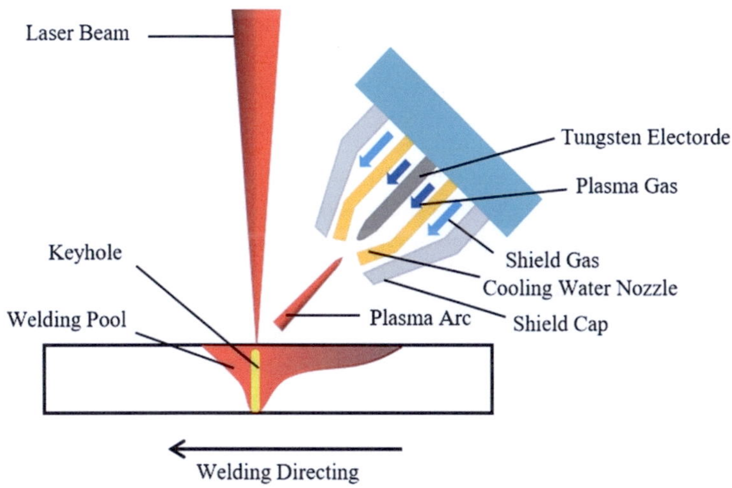

Fig. 2.3 Laser-PAW hybrid welding principal diagram

2.2 Physical Mechanism of Laser-Arc Hybrid Welding

Although the advantages of the hybrid welding process over those obtained from either arc or laser welding are well established, the fundamental physical processes are less well understood. Much of the existing literature detailing laser-arc hybrid welding is based on empirical studies related to the process capabilities [17]. In the following sections, the physical processes of laser-arc hybrid welding are more closely examined. Particular attention is paid to four important areas, including the physical effects of laser-arc welding, the effect of laser on arcs, the effect of arcs on laser, laser-arc interactions, and the characteristics of laser-arc welding.

2.2.1 Physical Effects of Laser-Arc Hybrid Welding

Laser-arc hybrid welding is a sophisticated joining technique that combines the advantages of laser and arc welding processes. This fusion welding method utilizes a laser beam in conjunction with an electric arc to create high-quality, high-strength welds. In this article, we will provide a comprehensive overview of the physical effects associated with laser-arc welding.

Heat generation and transfer: Laser-arc welding generates an intense heat source, primarily from the laser beam and the electric arc. The laser produces a localized, concentrated heat, while the electric arc contributes to the overall heat transfer. This unique combination allows for precise control over the welding process.

HAZ (Heat-Affected Zone) and thermal stress: The HAZ represents the region surrounding the weld joint that has been subjected to thermal cycling but has not reached the melting point. The laser-arc welding process minimizes the width of the HAZ due to the concentrated heat input, reducing potential distortion and stress on the workpiece. However, the rapid heating and cooling cycles may induce residual thermal stresses, which can impact the overall weld integrity.

Melting and solidification: As the heat is applied, the parent materials undergo melting, followed by a solidification process to form the weld joint. The laser and arc energy input significantly affect the melting and solidification rates, resulting in varying microstructural characteristics and mechanical properties of the joint.

Metallurgical transformations: During laser-arc welding, the material experiences various metallurgical transformations. The high-temperature environment alters the microstructure, phase composition, grain size, and potentially induces chemical reactions. These transformations influence the mechanical properties, such as hardness, tensile strength, and toughness.

Keyhole formation: The laser beam's intensity, combined with the electric arc, creates a high-energy density that can form a keyhole in the material. The keyhole is a vapor-filled cavity that facilitates deep penetration of the weld, allowing for efficient joining of thick materials. Keyhole stability and control are crucial for achieving consistent weld quality.

Plasma generation: Laser-arc welding generates a plasma plume due to the ionization of the vaporized material. The plasma plume can influence the energy distribution and shielding effectiveness during the welding process. Proper gas shielding is necessary to protect the molten pool and maintain stable arc characteristics [18].

Residual stresses and distortion: Due to the localized heating and cooling cycles, laser-arc welding can induce significant residual stresses and distortion in the welded structures. Understanding and managing these effects is essential for maintaining dimensional accuracy, especially in precision applications.

Laser-arc welding offers unique advantages in terms of heat control, deep penetration, and joint quality. However, it is crucial to consider and understand the physical effects associated with this welding process. By comprehensively studying the heat generation, metallurgical transformations, keyhole formation, plasma generation, and residual stresses, engineers and researchers can optimize laser-arc welding parameters and techniques for superior weld quality and structural integrity.

2.2.2 *Effect of Laser on Arc*

Since hybrid welding is often performed using a high-power laser operating in the keyhole mode, the laser energy is absorbed through multiple reflections of the laser in the keyhole. A plasma phase is also formed by the interaction between the high-power density of the laser heat source, the shielding gas and the metal vapor formed by the laser–material interaction. The plasma phase absorbs a portion of the laser energy, as described in detail in the section on "Absorption of arc and laser". Using a high ionization potential shielding gas, such as helium, the formation of laser induced plasma can be mitigated, particularly for high wavelength CO_2 lasers. During hybrid welding process, the interaction between the arc and the surrounding atmosphere also forms a plasma phase. The size of the arc plasma often depends on the distance between the heat sources and the arc current density, arc length and voltage. It has also been observed that the laser and arc plasma interact, and the transmission of the laser through the plasma results in some additional laser attenuation. In the case of Nd: YAG laser/GTAW hybrid welding of mild steel, the arc plasma absorbed up to 1% of the laser energy and the inverse Bremsstrahlung absorption coefficient was $0.01\ cm^{-1}$.

During hybrid welding, the arc undergoes a contraction in which its width is nearly the same size as the laser, because the presence of the laser induced plasma causes the arc resistance and radius to decrease. The influence of laser radiation on arc resistivity and stability can be explained by two phenomena. First, a small part of the laser energy is absorbed by the arc plasma, further ionizing the arc plasma and reducing its electrical resistance. Second, significant vaporization of the work piece material occurs at the location where the laser impinges on the surface of the work piece, and metal vapor is then transported into the arc plasma. Since the metal atoms have a much lower ionization potential than the shielding gas, the effective ionization potential of the plasma is reduced and a more conductive, stable plasma channel for the arc root and column is obtained. Since the arc follows the path of least electrical resistance, the arc tends to bend, and the arc root forms within close proximity to the keyhole.

2.2.3 *Effect of Arcs on Laser*

Laser-arc hybrid welding combines the advantages of laser and arc welding processes, offering improved weld quality and efficiency. The interaction between the laser and arcs in this technique plays a significant role in determining the welding performance. In this article, we provide a comprehensive overview of the effect of arcs on lasers in laser-arc hybrid welding.

Laser-arc interaction: The presence of arcs in laser-arc hybrid welding significantly influences the laser beam propagation and behavior. The interaction between the laser and arcs occurs through several processes, such as absorption, scattering, and plasma formation. Understanding these interactions is crucial for optimizing the welding conditions and achieving desired weld characteristics.

Absorption of laser energy: The arcs in laser-arc hybrid welding can absorb a portion of the laser energy, leading to reduced laser power available for material heating. The absorption characteristics of arcs depend on their properties, such as current, voltage, arc length, and plasma composition. Proper adjustment of laser parameters is necessary to compensate for the energy loss due to arc absorption.

Beam deflection and distortion: The arcs generate plasma and related thermal effects, resulting in beam deflection and distortion of the laser beam. These phenomena cause changes in the laser's trajectory and focus, affecting the heat input, penetration depth, and overall weld quality. Precise control of the laser beam path and adjustments to compensate for beam deflection are crucial.

Thermal interaction: The presence of arcs introduces additional heat sources in laser-arc hybrid welding. The thermal interaction between the arcs and the laser beam affects the temperature distribution and heat transfer in the welding zone. The complex interplay of laser and arc-generated heat influences the weld pool dynamics, solidification behavior, and microstructural development [19].

Keyhole stability and control: Laser-arc hybrid welding relies on the formation and stability of a keyhole for efficient deep penetration. The arcs contribute to keyhole formation by providing additional plasma and heat. However, improper arc conditions or instabilities can disrupt the keyhole and adversely affect the welding process. Optimizing the arc parameters and ensuring keyhole stability are essential for successful laser-arc hybrid welding.

Plasma generation and shielding: The arcs in laser-arc hybrid welding produce a plasma plume, which affects the welding process and shielding effectiveness. The plasma can influence the energy distribution, gas flow dynamics, and formation of the protective gas envelope around the weld pool. Understanding the plasma behavior and its interaction with the laser and shielding gases aids in achieving better process control and weld quality.

Process monitoring and control: The effect of arcs on lasers in laser-arc hybrid welding necessitates real-time monitoring and control systems. Monitoring techniques, such as spectroscopic analysis and imaging, can provide insights into the arc behavior, laser-arc interaction, and welding conditions. Feedback control systems can adjust laser and arc parameters dynamically, ensuring stable and optimized welding performance.

The effect of arcs on lasers in laser-arc hybrid welding is a critical aspect that influences the welding performance and weld quality. Understanding the laser-arc interaction, absorption of laser energy, beam deflection, thermal interaction, keyhole stability, plasma generation, and process monitoring is crucial for optimizing the welding parameters and achieving consistent and high-quality welds. Further research and advancements in laser-arc hybrid welding technology will continue to enhance the understanding and control of this interaction, leading to improved welding processes and applications.

2.2.4 Role of Laser-Arc Separation

The roles that both heat sources play in these synergistic interactions are very significant. Often, hybrid welding is described as simply welding with an arc and a laser. However, changes in the distance between the arc and laser sources can cause two distinct welding processes, hybrid and tandem welding, to emerge. During hybrid welding, an interaction occurs between the laser and arc plasmas as a result of the close proximity of the two heat sources. Tandem welding occurs when the distance between the heat sources is significantly greater than the arc plasma radius, usually 5–8 mm. In tandem welding, the arc and laser act separately on the work piece.

The distance between the heat source focal points plays a large role in the arc and laser induced plasma interactions. Experimental research has shown that during the plasma interaction, the contracted arc rooting in close proximity of the keyhole can increase weld pool penetration depth. This increase in depth is small relative to the penetration depth achieved by laser welding alone. In addition, the interaction results in a significant increase in weld pool width over laser welding. Beyond the fact that these weld pool dimensions increase due to the interaction of the plasmas, little scientific explanation as to the fundamental physical processes which occur to enhance these weld pool dimensions has been presented.

The exact role of inter-heat source spacing on the interaction of the heat sources and its effects on the physical processes is not well defined and requires further research. When the distance between the laser and arc is significantly greater than the arc plasma radius, the laser and arc plasmas are separate. In contrast, when the distance between the two heat sources is less than or approximately the same as the arc plasma radius, the two plasmas interact. Therefore, the interaction of the laser beams and the arc plasma mainly depends on the distance between the heat sources, arc radius and arc plasma radius. A quantitative understanding of the inter heat source spacing and its relationship with welding parameters will lead to a significant improvement in process control.

Recent research has examined the importance of the order in which the laser and arc are arranged during hybrid welding. For example, during hybrid welding of mild and stainless steels, the laser is placed in front of the arc in what is known as backhand welding. The main benefit of the arc following the laser is to decrease the high temperature gradient developed by the laser, resulting in a decreased propensity for solidification cracking and an increased toughness. On the other hand, it is common during the welding of copper and other alloys with high reflectivity or thermal conductivity for the torch to be placed before the laser (forehand welding) in order to preheat the material before the laser passes.

The advantage of having the arc come before the laser is to increase both the amount of laser energy absorbed by the material and the efficiency of the laser energy input. However, experimental work performed on 304 stainless steel using Nd: YAG laser/GTAW hybrid welding has shown that the order of the heat sources has no clear effect on the depth of the weld. A certain distance between the point of the arc and the laser beam needs to be maintained to ensure weld quality and that the interaction of the laser and arc plasmas is maintained. Naito et al. [20] showed that, when the arc is leading, hybrid welding does not always result in a greater weld pool depth, but the increase in depth is primarily dependent upon the distance between the heat sources. Further research is required on the role of heat source separation and its effect on plasma interactions, plasma formation and surface heating interactions of the heat sources.

2.2.5 Spectral Characteristics of Laser-Arc Welding

Laser-arc welding, a combination of laser and arc welding processes, is a versatile fusion welding technique that offers enhanced precision and high-quality welds. The spectral characteristics of this welding method play a crucial role in understanding the underlying physical processes and optimizing the welding parameters. In this section, we provide a comprehensive overview of the spectral characteristics associated with laser-arc welding.

Plasma emission spectroscopy: Plasma emission spectroscopy involves analyzing the emitted light from the welding plasma to investigate the elemental composition and temperature of the plasma plume. The spectroscopic analysis of plasma emission lines provides valuable information about the ionization states, excitation, and thermal energy distribution within the welding process.

Laser beam spectral characteristics: The laser beam used in laser-arc welding exhibits specific spectral characteristics, which impact the welding process. The laser's wavelength, intensity, and beam profile influence the energy absorption and penetration depth into the material, affecting the overall quality and efficiency of the weld.

Arc spectral features: The electric arc in laser-arc welding produces its characteristic spectral features, mainly in the ultraviolet, visible, and infrared regions. Analyzing the arc's spectral emission lines helps in understanding the arc stability, plasma characteristics, and heat generation within the welding process. Spectroscopic investigations enable the control and optimization of arc parameters to achieve desirable welding results.

Spectral imaging: Spectral imaging techniques, such as hyperspectral imaging, provide a comprehensive visualization of the spectral information across the entire wavelength range. These techniques offer valuable insights into the spatial distribution of various spectral components, plasma characteristics, and temperature distribution within the welding zone. Spectral imaging enables real-time monitoring and control of the welding process for enhanced quality and consistency.

Absorption and reflection spectra: The optical properties of the materials being welded influence the spectral characteristics of laser-arc welding. The absorption and reflection spectra of the materials determine the energy absorption and transmission in the welding process. Understanding these characteristics helps in selecting appropriate laser parameters and optimizing the welding conditions for different material combinations [21].

Spectral signature analysis: Spectral signature analysis involves the characterization and identification of specific spectral features associated with different welding conditions, materials, and defects. By analyzing the spectral signature, such as peak intensity, wavelength shifts, and line broadening, researchers can develop diagnostic tools for process monitoring, defect detection, and quality control in laser-arc welding.

The spectral characteristics of laser-arc welding provide valuable insights into the underlying physical processes and enable optimization of welding parameters for superior weld quality. Through plasma emission spectroscopy, analysis of laser and arc spectral features, spectral imaging, absorption/reflection spectra, and spectral signature analysis, researchers can improve their understanding of the welding process, monitor real-time changes, and develop effective control strategies. This knowledge contributes to advancements in laser-arc welding technology and its wide-ranging applications in various industries.

2.3 Molten Pool Flow and Dynamic Behavior of Keyhole

The laser-GMAW hybrid welding process includes a variety of complicated physical phenomena such as mass transfer, heat transfer, fluid flow and phase interaction [22]. The transient behavior of the keyhole can be divided into three characteristic stages, linear growth stage, oscillation growth stage and high frequency oscillation stage. The first stage lasted for a short period of time only about 6 ms. In this stage, the temperature of the workpiece surface rose rapidly to the boiling point under the influence of direct laser irradiation. In such case, a large amount of metal vapor was produced and the depth of the keyhole changed rapidly. In the second stage, the value of keyhole depth gradually increased to the maximum, while the growth rate of the keyhole gradually decreased. During the process of increasing the keyhole depth, surface tension and hydrostatic pressure gradually increased. The variation of these forces hindered the increase of keyhole depth. In addition, the process of increasing keyhole depth was accompanied with periodic oscillation due to the periodic fluctuation of keyhole wall that affected the laser absorption rate of base metal. In the third stage, the keyhole depth remained relatively stable and changed slightly. It was considered that the driving force such as recoil pressure was basically equal to the hindering force such as surface tension and hydrostatic pressure. However, the factors affected the stability of the keyhole were very complicated. The value of such forces fluctuated within a certain range with the change of keyhole shape and fluid flow state, in which case the keyhole was in dynamic equilibrium.

2.4 Control of Weld Formation and Porosity in Laser-Arc Welding

2.4.1 Weld Formation

A large number of process parameters are to be controlled during the laser-arc hybrid welding process. Hybridization of the processes requires new parameters to be defined in addition to the control parameters of the individual units. The optimum process parameters of the laser-arc hybrid welding process cannot simply be determined from the combination of optimum parameters of the individual welding processes. The optimum value of the parameters may change due to the interaction effect of the laser beam with the electric arc. For example, the laser power requirement for laser-arc hybrid welding is less than that for the autogenous laser welding process, due to the use of a secondary heat source [23].

(1) **Laser power**

The laser power is the main source of heat during hybrid laser arc welding, which produces the keyhole and ensures the deep penetration welding. Because of using electric arc with the laser at the same weld pool, the laser absorption within the metal and the circulation of molten metal at the weld pool are increased. The power density has the greatest effect on welding during welding with laser. The weld penetration increases with laser power. It is observed during hybrid laser-GMAW welding of AZ31 magnesium alloys that the weld penetration increases 2–3 mm/kW of laser power, as compared to the increase of 0.5 mm/kW of arc power. The results of bead-on-plate hybrid laser-GMAW of A5052 aluminum alloys and SUS304 stainless steel with fiber laser using laser power ranging from 2 to 10 kW, showed the increase of weld penetration with laser power for both the alloys. It is observed during hybrid laser- GMAW of copper alloy that the face width, root width and root reinforcement height increase, and the face reinforcement height decrease with the increase of laser power. It is also found that laser absorption increases when the keyhole is formed at the weld zone due to the application of high power density [24]. Welds with higher joint strength without obvious defects can be obtained at favorable laser power range.

(2) **Welding speed**

Higher welding speed is the characteristic feature of hybrid laser arc welding, which increases productivity. However, weld width and weld penetration are inversely affected by the welding speed. To compensate for this, more laser power is required to be applied and the filler wire feed rate to be increased to maintain a constant volume of weld pool. The welding speed to filler wire feeding ratio is an important factor for the stability of the keyhole and for the entire process. A too high welding speed may cause improper fusion or lack of penetration of the weld, whereas, a too low welding speed may create a larger weld pool with deep penetration till the limit of burn through the material and may cause partial decomposition. The fast heating and cooling cycle due to higher welding speed produce steep temperature gradients at the weld zone, which may result in metallurgical defects within weldment. However, an increase in welding speed can effectively reduce the thermal residual stress concentration.

(3) **Relative positioning of the laser beam and the arc welding torch**

The relative positioning of the laser beams and the arc welding torch is also a factor. The laser beam is always directed perpendicular to the welding direction to get the maximum possible weld penetration. The arc welding torch can be positioned in front of the laser beam (arc leading) or just behind the laser beam (laser leading). During the arc-leading process, the torch is positioned at a drag angle. During the laser-leading process, the welding torch is positioned at a push angle. Dragging generally produces deeper penetration and a narrower weld width than the pushing technique. In an arc-leading process, the weld deposition is narrow and convex as compared to the laser leading process. The effect of leading or trailing arrangements of the arc welding torch was found to be quite insignificant. Only an improvement of 10% weld penetration is observed when the MAG torch is placed behind the CO_2 laser beam. Contrary to this, an improvement in weld penetration is observed when using arc leading configuration during the welding of type 304 stainless steel using the laser-GTAW hybrid welding process with a 1.8 kW Nd: YAG laser.

(4) **Laser to electrode separation distance**

The arc electrode must be in close proximity to the laser beam, to ensure their synergic effect in the common weld pool. The distance between laser and arc significantly affects the arc characteristic, droplet transfer mode and weld bead geometry. The laser to electrode separation distance is typically kept from 0 mm (for co-axial arrangement) to 5 mm (for paraxial arrangements), depending on the types of laser parameters, arc welding parameters, and materials used. Increasing the gap between the laser beam and the electrode may result in loss of the hybridization effects. Co-axial arrangement or a very short gap may lead to a problem of absorption of laser energy by the arc plasma, which partially blocks the laser beam resulting in lesser penetration. The laser to arc distance affects the morphological characteristics of the weld pool. The weld pools appear more asymmetrical with the increase of lateral displacement of the GMAW torch from the fixed laser source. It is observed that increasing the laser to arc distance, changes the weld bead geometry from "cocktail cup" to "cone-shaped".

(5) **Positioning of focal point**

The focal position of the laser beam affects the penetration of the weld. The maximum weld penetration is generally achieved when the focal point of the laser beam lies below the surface of the workpiece. The focus of the laser beam changes when it penetrates the molten pool. The volume of the weld pool in laser-arc hybrid welding is much greater than autogenous laser welding. The formation of cup-shaped weld deposition in laser-arc hybrid welding increases the curvature of the top surface of the melt pool, which changes the focus of the interacting laser beam. The focus of the laser beam during CO_2 laser-GMAW hybrid welding was reduced by 0.7 mm versus autogenous laser welding. The optimum focus of the laser beam is shifted from − 2 to − 4 mm during the welding of a mild steel plate using CO_2 laser-GMAW hybrid welding with a 0.6 m/min welding speed.

(6) **Electrode angle**

The arc electrode angle is provided in such a way that the shielding gas coming out from the torch nozzle effectively deflects the laser induced plasma from the weld zone to increase the laser penetration of the weld. The electrode angle is typically set around 45–65° from the workpiece surface, which reduces the arc length. The laser beam is generally directed normal to the workpiece surface to obtain better penetration. However, the laser beam is tilted to an angle during the welding of highly reflective materials to avoid any damage to the laser head due to the reflected beam.

(7) **Shielding gas composition**

The shielding gas is generally used during welding to protect the weld zone from the surrounding atmosphere. The use of shielding gas also influences the arc characteristics, the formation of weld profile, and the mode of metal transfer. Inert gases such as helium and argon are the predominant constituents of shielding gases used in laser-arc hybrid welding processes. Nitrogen, carbon dioxide, and carbon monoxide (CO) are also used as shielding gas constituents based on the selection of materials, the laser source, and the arc parameters. The arc stability and the plasma shape depend on the composition of the shielding gas used in laser-arc hybrid welding, which affects weld penetration. The laser energy absorption by the laser-induced plasma or by gas plume during the CO_2 laser-arc hybrid welding process is a well-known problem. This reduces the laser intensity reaching the weld pool, decreasing the weld penetration. The use of a high ionization potential shielding gas such as helium during welding with a CO_2 laser reduces the effect of plasma absorption. A mixture of argon, helium, and CO_2 is also used in the CO_2 laser-arc hybrid welding process. The use of helium ensures deeper penetration, whereas the use of argon improves the arc stability. In CO_2 laser welding, the optimum percentage of helium proportion varies from 30 to 80%, depending on other parameters and the material used. It is found during welding of high-strength steel using the CO_2 laser-GMAW hybrid welding process, that using helium at less than 30% in shielding gas requires an additional source of helium as a secondary plume suppression gas. The use of CO_2 in shielding gas up to 10%, during the CO_2 laser- GMAW hybrid welding process, is found to stabilize the arc. The use of a small percentage of oxygen ($\leq 5\%$) in the shielding gas reduces the spatter formation and improves the metal transfer during welding. However, no significant vapor plume absorption of laser energy is noticed during the welding with lasers having a shorter wavelength such as Nd: YAG lasers, fiber lasers, disc lasers, etc. Thus, the choice of shielding gas composition during welding with Nd: YAG, fiber, and disc lasers is based mainly on the requirement of shielding and arc stability. The shielding gas delivered through the arc welding torch is generally sufficient to protect the molten weld pool. However, an additional nozzle/torch is used in some cases, such as welding of reactive metals like titanium, to supply shielding gas, which trails the arc. This provides a shielding atmosphere during the cooling to avoid the formation of oxides.

(8) **Power modulation of the arc welding system**

Due to the dynamic behavior of the arc, the current and voltage change continuously during the arc welding. The arc power is controlled by the welding current and voltage. Current positively affects the rate of melting of the consumable electrode, filler and the base materials. Voltage affects the arc length and the welding current. A constant current power source is generally used for the GTAW or PAW, whereas, a constant voltage power source is used for the GMAW process. In GTAW or PAW, the current is preset, and any changes of welding current due to the change of arc length is not critical because of the use of non-consumable electrode and separately used filler material. In GMAW, the voltage and wire feed rate are preset, and the welding current is controlled by the wire feed speed. The change of wire feed speed changes the arc length and arc voltage, which it turn change the arc current and the melting rate of the electrode to maintain an equilibrium arc length.

Conventional GTAW is generally carried out with an AC power source, because it prevents the electrode from overheating, removes oxides during the positive phase and heats up the work piece deeply during the negative phase. AC power source is also used in hybrid laser GTAW process. The use of DC power source in hybrid CO_2 laser GTAW is resulting in higher energy input and longer electrode life. The DC power source with electrode positive is used in conventional GMAW process as well as in hybrid laser-GMAW process, which increases the deposition rate. The use of pulse mode in both, hybrid laser-GTAW or hybrid laser-GMAW processes, reduces the amount of spatter and heat affected zone. Pulse width and pulse frequency are the additional parameters to be controlled during use of pulsed current. Pulse width determines the time duration of the pulse (peak current), and also affects the size of metal droplet and the width of arc cone. The pulse frequency controls the overall heat input to the weld zone. Pulsed arc welding in combination with pulsed laser can also be utilized for hybrid laser arc welding.

The mode of metal transfer is an important factor which is affected by the arc power, electrode stick-out and electrode diameter, laser to arc distance, and shielding gas composition and gas pressure. There are three different modes of metal transfer observed during arc welding are projected/spray transfer, gravitational/ globular transfer and short circuit transfer. The magnitudes and directions of electromagnetic force and plasma drag force acting on the droplet affect the behavior of droplets. The temperature field, fluid flow pattern and geometry of weld pool are deeply affected by the droplet impingement process, due to the transfer of mass, energy and momentum into the weld pool. Spray transfer is much preferred in hybrid laser arc welding to ensure deep penetration due to high heat input per unit length during maintaining high welding speed. Lower turbulence of weld puddle is observed during use of spray mode metal transfer due to projection of small filler droplets into the weld pool. The use of pulsed current enhances the spray metal mode transfer because the use of peak current with a small pulse duration and high pulse frequency affects the formation of metal droplets and detach them frequently. This reduction in the chance of formation of bigger droplets is a characteristic feature of gravitational transfer. Globular metal transfer mode is observed at a small laser to electrode separation distance while a spray mode is achieved with an increase in the separation distance.

The ratio of laser power to arc power is also an important factor for hybrid laser arc welding which affects the weld geometry, metallurgical property of the weld, residual stress and deformation pattern. This balances the effect of the laser beam and arc on the common weld pool. Higher energy ratio ensures narrow and deep weld, and increases the welding speed. The weld bead size increases with arc power, till steady state condition achieved. However, very high arc power causes a reduction in penetration depth because of the change of metal mode transfer from spray transfer to globular transfer.

(9) **Wire feed rate**

The feed rate of the consumable electrode is determined based on the requirement of metal deposition at the weld. In a constant voltage GMAW setup, the wire feed rate is preset. The higher wire feed rate enables the operator to join a weld with a higher part gap, also helping to increase the welding speed. The selection of wire feed rate is critical to avoid a lack of deposition or the supply of extra material. With a higher wire feed rate, the higher current is required to increase the rate of deposition. The increase of the wire diameter also requires higher current to maintain the constant arc length.

(10) **Joint gap**

The laser beam can weld parts with a gap up to 0.2–0.25 mm only, without visible defects like undercut or incomplete weld bead. However, a joint gap of 1 mm can easily be welded with hybrid laser arc welding process. The laser to arc energy ratio, laser arc distance, welding speed and wire feed rate are the critical parameters which determines the rate of deposition of the filler materials to fill the joint gap. It is observed during the hybrid CO_2 laser-GMAW welding of high strength carbon steel that the wire feed rate is to be increased with the joint gap. The results presented in

Ref, shows that the welding speed can be increased during welding of parts having joint gap up to 1 mm, because the small joint gap simplifies the penetration, as more filler material reaches to the root of the joint. However, increasing the joint gap more than 1 mm requires the welding speed to be reduced and also requires proper selection of process parameters. A larger joint gap may require multiple weld passes and the use of backing plate.

(11) **Joint configuration and edge preparation**

The basic joint configurations used in welding are butt, corner, edge, lap, and tee joints. All these joint configurations are used in the laser-arc hybrid welding process. Due to the small spot diameter of the laser beam, the square edges are used for welding, whereas grooves are made at joints when using arc welding, especially for thick sections. For thick materials welding using arc welding V, U, or J grooves are made at joint edges to ensure that the molten filler material can reach the root of the joint. The shape of the groove determines the volume of the materials deposition and the number of weld passes required, which influences joint strength, residual stress distribution, and bead geometry. For heavy sections, double V, double U, or double J grooves are used which reduces the volume of the materials deposition required, produces symmetric welds, and minimizes warpage. The simultaneous action of joint filling by filler deposition using arc welding and deep penetration by laser requires a different joint design for laser-arc hybrid welding. Thin metal sheets or metal plates having a thickness less than 6–8 mm do not require edge preparation for welding using laser-arc hybrid welding. For material having a thickness more than 6–8 mm, preparation of edges is required. The grooves used in parts for laser-arc hybrid welding generally are smaller, including angle, narrow root opening, and larger land thickness as compared to the grooves used in arc welding. The V groove used for laser-arc hybrid welding resembles the shape of a wine glass or a funnel. For a part having a different thickness, the groove is made in such a way that the land thickness of the part should be the same at the joint while the height difference is to be compensated at the cap. The large weld pool at the cap has better tolerance to geometrical mismatch [25].

2.4.2 *Weld Porosity*

Porosity formation and mitigation in LAHW is more complicated since two sources are introduced in welding. Identically to autogenous LBW, there are two porosity formation mechanisms: (i) keyhole collapse and (ii) unfavourable melt flows. The latter is affected by both the laser beam and the arc. The situation is more complex when an MIG source is used. In LAHW, the wire applied may also influence the porosity through the presence of surface flaws and cracks containing dirt and moisture resulting in hydrogen gas pores.

(1) **Laser and arc power**

According to Katayama et al., LAHW can mitigate porosity formation using increased arc current from 60 to 240 A [26]. No porosity was found using 240 A for 3.0 mm AA5052. Bubbles filled with vapour are generated from the bottom part of the keyhole due to an intense evaporation from the front wall causing collapse of a deep vapour capillary. The collapse is generated due to unstable flows of molten metal and bubbles are trapped by the solidification front. At 240 A, the surface of the molten pool is concavely depressed due to the higher arc pressure providing faster melt flows, thus the keyhole diameter is much larger. At the same time, the penetration depth was significantly increased with the width of the fusion zone.

The balance of laser/arc power ratio was studied by Leo et al. where 3.0 mm AA5754-H111 was joined with fiber laser-MIG with filler wire containing various levels of Mg content [27]. It was found that an increase of arc power also suppressed the porosity, providing better tensile strength with lower Mg evaporation. Improved performance was at slightly higher arc power ratio to laser beam power. Currents that were too high provided slightly larger grain size with porosity. However, the weld strength was much lower than the BM, exhibiting only 68%. Higher laser power can benefit to higher stability of the keyhole by providing less porosity. As a result, laser/arc power is not straightforward and requires balancing depending on other welding parameters, e.g., welding speed. Porosity may increase with increase of laser power (from 2.5 to 4.0 kW) in case of 6.0 mm AA2024-T3 due to keyhole collapse. By variation in the welding speed (from 1.0 to 2.5 m/min at 3.0 kW) at lower laser power, no pores were generated since the process was in heat conduction mode with wide weld geometry at shallower penetration depth. As an alternative, a defocussed laser beam may be applied to reduce porosity.

Based on results evaluated here, the laser/arc balance is intricate to optimize since other parameters should be also considered, e.g., the welding speed. In general, the best way is to achieve keyhole stability by balancing welding parameters. Another suggestion is switching from keyhole mode towards transition mode or even to heat conduction mode. Lower depth of the keyhole reduces collapsing at the bottom. This is somewhat contradictory. However, it may enable high welding speeds, which surpass the conventional arc welding speed by several times. Moreover, some balancing of the parameters, e.g., increase of arc current and appropriate process distance, allows not only to increase the laser beam power at fast welding speeds but also to enhance penetration depths.

(2) **Shielding gas**

The effect of He/Ar shielding gases is not widely published in detail. Pure argon shielding gas is the most common for Al alloys due to cheapness. However, due to unstable keyhole processing, He–Ar mix shielding gas can be considered as alternative since LAHW is cost-efficient process compared to conventional arc. Ar–He mixed gas may be a viable option. High ionisation potential of the mixed gas reduces the plasma size and density leading to higher penetration depths but can provide unstable arc conditions in pure He or even by adding more than 40% of He. Tani et al. suggested that Ar is safer to use due to its heavier mass with more reliable protection without a need to use a high flow rate compared to He, which is lighter and may not properly cover the weld pool [28]. Shielding gas flow rate should not be high to avoid turbulent distribution providing unstable processing and optimal shielding of melt pool. More interesting is that the angle should be horizontally flat (at 0°) or inclined up to 30°. In LAHW this angle is always much higher and thus inherently possess non-optimal shielding. In such a case, additional shielding can be applied, e.g., a special nozzle or a back extended diffuser, for the tail of melt pool seems to be reasonable to protect from oxidation. The application of pure nitrogen reduces porosity and is a viable option for LBW due to its cheapness and for LAHW if the arc is stable.

(3) **Arc position**

Arc position relative to the laser beam may have substantial effect on generation of the porosity but only few studies are reported on this issue. Trailing arc setup seems much more effective in suppressing porosity due to better stability since the droplets frequently hit the keyhole opening area in leading arc.

(4) **Process distance**

Katayama et al. identified that the maximum penetration depth was at 2.0 mm DLA with trailing arc setup. At 0 mm, droplet impinged the laser beam opening area and disturbed the process [29]. The effect of DLA on porosity was not clearly identified, but it was stated that lower porosity was in full penetration welds compared to partial penetration. Thus, optimal DLA to avoid porosity should be within the range of 2.0 to 5.0 mm. At certain process distances, one or another arc position can provide different results due to strong interaction. It is expected, that in the leading arc position, an increase in the process distance may provide acceptable quality of welds with less porosity and more stability. This can be especially true in the case of thicker plates.

(5) **Welding Speed**

The welding speed in LAHW is substantially more intricate than for LBW, where it is basically unrestrained and can be up to 10 to 15 m/min. Due to the added arc, there are restrictions in use of high welding speeds (> 2.0 m/min) since the arc rooting and subsequently the stability is sharply becoming unstable. It is reported that with increasing welding speed, the porosity formation tendency is reduced similarly to LBW. Controversially, Han et al. reported higher porosity with increased welding speed (from 0.7 to > 1.2 m/min) related to more rapid solidification and narrower welds preventing bubbles from escaping the weld pool [30]. As a result, there should be a delicate balance between laser beam power, arc power, welding speed, and other associated parameters.

2.5 Applications for Laser-Arc Welding

2.5.1 Application of Laser-Arc Hybrid Welding in Shipbuilding Industry

Wide range of welding processes involving laser as one of the energy sources are under active development. Majority of the welding work in industries dealing with sheet metal joining is still carried out using GMAW, however switching to laser-arc hybrid welding would bring significant economic benefits. Laser-arc hybrid welding could be suitable for large number of serial production applications where throughput volumes are high. In laser-arc hybrid welding, the beam and the arc act simultaneously, support each other and in comparison, to laser welding produce improved gap bridge ability and in comparison, to arc welding higher processing speeds, increased quality, less distortion and thus less re-work and shorter production times. It provides combination of "properties" that usually would be mutually exclusive. Typically, due to laser beam penetration ability and synergy of the processes, only one pass is required to form a solid joint. Additionally, the workpieces are subjected to smaller heat input, which reduces distortions and the need for post welding corrections, thus saving time and money. The reasons of laser-arc hybrid welding not being more widespread is that it is seen as a complex process with high number of variables involved and expensive initial investment. However, many process parameters can be changed and still a sound weld can be deposited, because the quality window of laser-arc hybrid welding is significantly larger than that of autogenous laser welding.

As shown in Fig. 2.4: EH36 is welded by laser-arc hybrid welding. The weld appearance is formed at different speeds, the partially penetrated laser-arc hybrid welding welds overlap appropriately, resulting in a completely closed weld cross section. The areas of applications of structural steel range from assemblies on land, heavy industry vehicles to offshore platforms and power plants. The final product is often subjected to both, static and dynamic loading and good weldability is essential for safety and long-life cycle of the components. Laser and laser-arc hybrid welding processes deliver reduced amount of heat to the joint and use less, if any, consumables to form a solid connection, therefore having both, metallurgical and economic benefits, most notable being the greatly reduced costs on the post-weld straightening and correction work. Both of the processes are very versatile and automatable, which contributes to uniformity of the welds produced and high repeatability of the process.

Autogenous laser welding has had vigilant reception by industry regardless of aforementioned benefits. The small amount of adjustable process variables simplifies finding the optimal process parameter window. Due to small dimensions of the tightly focused laser beam spot on the surface, the factory floor solutions are at times unable to comply with extremely precise part positioning requirements, especially when the welds are long and joint gap is changing because of metal heat expansion or slight distortions which might occur during the welding. On the other hand, in automotive and other industries where material thickness is small and lap, butt and eccentric fillet joints are common, benefits of autogenous process are the greatest. Regardless

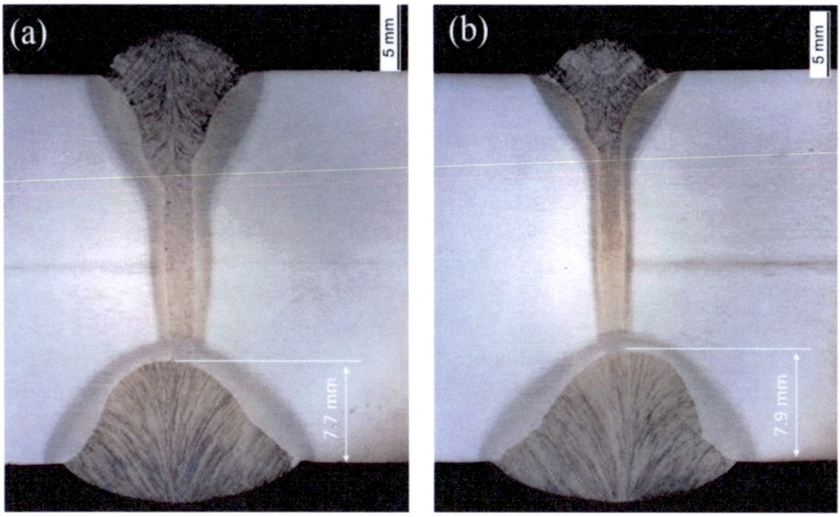

Fig. 2.4 The weld appearance of laser-arc hybrid welding at different welding speeds is **a** 0.6 m/min **b** 1.15 m/min [31]

of relatively simple setup and vast savings in material (joint is formed from the workpiece material without additional fillers), low gap bridging capacity and small parameter window for acceptable quality welds are seriously hindering wider use of autogenous laser welding in production of structures with T-butt and fillet joints, the risk of the beam misalignment from the joint being the key problem.

In laser-arc hybrid welding process, the energy to form a weld is delivered by two sources acting in synergy. Laser beam is responsible for penetration depth and high welding speed, while arc process allows bridging wider gaps and defining the shape of the bead. GMAW delivers additional heat that slows down the cooling and helps to avoid the formation of brittle microstructures. The molten pool near the surface is dominated by the arc process, while in autogenous laser welding the molten pool geometry determines the solidification process, which, in turn, is determined by the keyhole mode. Laser beam stabilizes the arc, and the processing speed is significantly higher than in arc or laser welding alone. Additionally, the spray transfer of filler material into the melt pool is beneficial for the mechanical properties of the joint. The hardness values in laser-arc hybrid welding are lower than those seen in autogenous laser welds, typically not exceeding the 350 HV hardness limit set by the standard for structural steels. Flexible beam delivery and short wavelength of fiber laser have eliminated for example the difficulty of choosing the shielding gas, which is an issue with CO_2 laser-arc hybrid welding. The wavelength of the fiber laser is not prone to creating plasma and therefore the shielding gas choice is based purely on the arc process specification. Study on bead formation in autogenous laser welding of butt and corner joints proposed a matrix flow chart to display the geometries and occurring details systematically.

2.5.2 Application of Laser-Arc Hybrid Welding in the Field of Construction Machinery

The crane outrigger is usually made of high-strength steel. Laser-arc hybrid welding can achieve high temperature and high energy welding, ensure the strength of the welding seam to meet the requirements, and provide excellent welding strength and toughness. The crane boom has an important influence on the weight and load of the whole machine. Laser-arc hybrid welding can obtain high-quality welded joints, reduce the volume and weight of the welded joints, and thus reduce the weight of the overall structure. The crane boom has an important influence on the weight and load of the whole machine. The heat input of laser welding is relatively small, which can control the heat affected range of the welding area, reduce the risk of deformation and deterioration, and ensure the geometric shape and dimensional accuracy of the outrigger. Laser-arc hybrid welding can be integrated with automation equipment and robots to achieve efficient and continuous production of crane outriggers and improve production efficiency and consistency. Jacek Górka et al. [32] used laser-arc hybrid welding to weld S700MC steel, which is widely used in engineering field. The macroscopic morphology is shown in Figs. 2.5 and 2.6, no welding defects were found, such as cracks, pores, incomplete fusion or incomplete penetration.

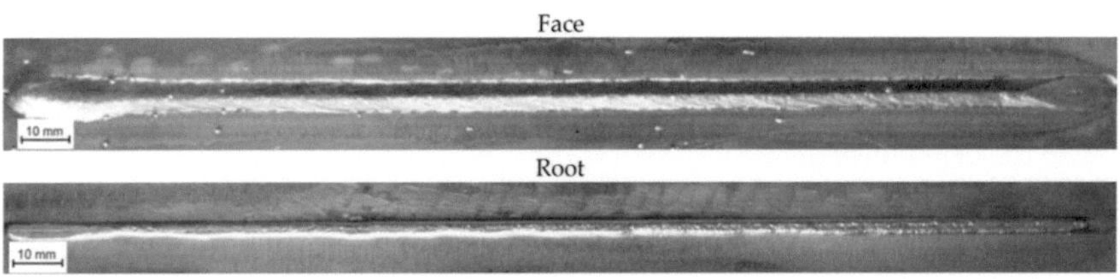

Fig. 2.5 The face and the root of the laser-arc hybrid welding joint in steel S700MC [32]

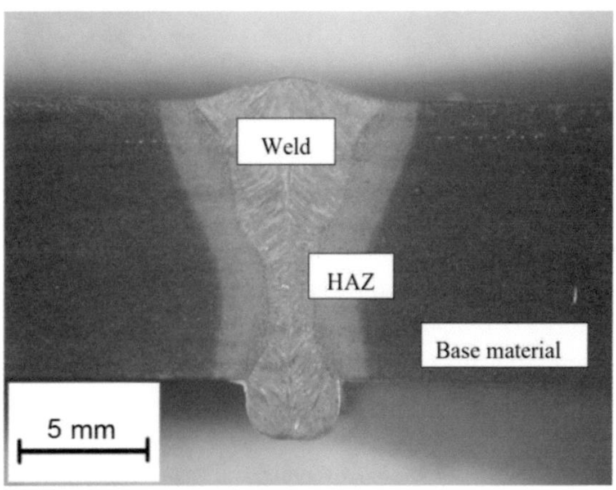

Fig. 2.6 Macrostructure of the laser-arc hybrid welding joint made of steel S700MC [32]

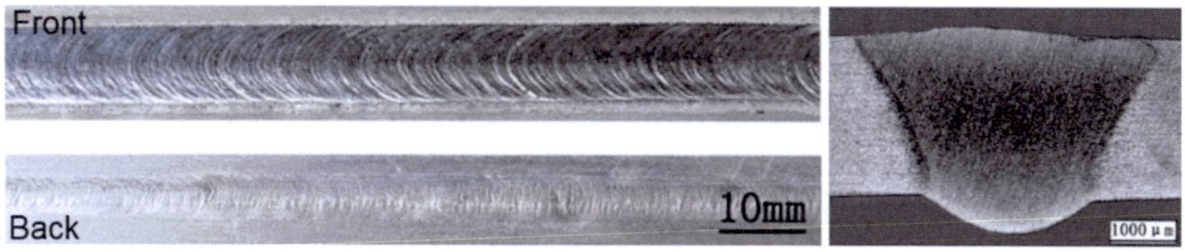

Fig. 2.7 Appearance and the macrostructure of laser-arc hybrid welding Al alloys joint [33]

2.5.3 Laser-Arc Hybrid Welding in the Field of Transportation

Laser-arc hybrid welding combines the advantages of laser welding and arc welding, which can achieve higher welding speed. This is very important for the production of large-scale subway car bodies, which can improve production efficiency and speed up delivery time. Subway car body is usually made of thin steel plate, and laser welding can realize welding under small heat input, reducing deformation and heat influence, so as to improve the welding quality of the car body. Laser-arc hybrid welding has significant advantages in the welding of subway car body, which can improve the welding quality, accelerate the production speed, save the cost, and meet the needs of thin plate welding of subway car body. Therefore, it is widely used in the manufacturing and production process of subway car bodies.

Hongyang Wang et al. [33] used laser-arc hybrid welding to weld 3 mm thick aluminum alloy for the application in the automotive industry (shown in Fig. 2.7). The welding appearance is uniform and there is no hump defect, welding pores or cracks in the joint. For the new generation of high-speed trains, the traditional arc welding process is difficult to meet the requirements of precision welding deformation control, improvement of service safety and reliability. Harbin Welding Research Institute Co, in collaboration with Qingdao Sifang Locomotive Co, has successfully applied laser-arc hybrid welding technology to the production of new generation of high-speed train. The floor, side wall and ceiling of the new high-speed train are all welded using laser arc hybrid welding process. Compared with the traditional arc welding, the welding efficiency is increased by about 4 times, and the welding deformation is reduced by 70% -90%.

2.6 Summary and Prospects

Laser-arc hybrid welding, as an advanced welding technique that combines the power of laser and arc, has brought revolutionary changes to the welding industry. Over the past few years, laser-arc hybrid welding has received extensive research and application, resulting in significant advancements. Looking ahead, this technology will continue to evolve and provide innovative solutions to welding requirements across various sectors.

Improvement of welding quality and efficiency: Laser-arc hybrid welding technology has the characteristics of high energy density and precision welding control, which can achieve high quality welding. With further breakthroughs and improvements in technology, laser-arc hybrid welding will be able to achieve deeper, narrower and more stable welds, thereby improving welding quality. At the same time, by optimizing welding parameters and improving equipment, laser-arc hybrid welding technology is expected to achieve higher welding speed and efficiency.

Expanding application areas: Laser-arc hybrid welding technology is suitable for welding a variety of materials, including metal and non-metal. With the in-depth study of welding processes of different materials, laser-arc hybrid welding will be able to be applied to a wider range of fields. Whether it is automotive manufacturing, aerospace, electronic equipment or medical equipment, laser-arc hybrid welding will provide innovative solutions for these industries to meet the needs of high-quality welding in different fields.

Achieviment of complex weld shape: With the help of advanced control systems and machine learning algorithms, laser-arc hybrid welding can achieve a higher degree of automation and accuracy to cope with various shapes and more complex welding tasks.

As an important innovation in the field of welding technology, laser-arc hybrid welding shows its broad application prospects and unlimited development potential. We can expect this technology to make greater breakthroughs and progress in welding quality, efficiency, application fields and welding shapes. At the same time, we also need to pay close attention to safety and environmental protection issues, strengthen research and training, and make efforts to promote the development of laser-arc hybrid welding technology to help the welding industry move towards a more prosperous future.

Problems

1. What is laser-arc hybrid welding? How does it differ from traditional welding methods?
2. List and explain the advantages and application areas of laser-arc hybrid welding.
3. Describe in detail the working principle and welding process of laser-arc hybrid welding.
4. Which types of materials are suitable for laser-arc hybrid welding? Why?
5. What are the key factors for adjusting the welding parameters in laser-arc hybrid welding? How can they be optimized?
6. Analyze practical case studies of laser-arc hybrid welding in shipbuilding.
7. Discuss the advantages and challenges of laser-arc hybrid welding in ship structure welding.
8. Explain the impact of laser-arc hybrid welding on the mechanical properties of welded joints.

9. Discuss the advantages and limitations of laser-arc hybrid welding in ship plate repair.
10. Investigate the application effectiveness of laser-arc hybrid welding in ship coating welding.
11. Compare laser-arc hybrid welding with other ship welding methods such as arc welding and laser welding.
12. Explore the automation and industrial application of laser-arc hybrid welding in the shipbuilding process.
13. Explain the contribution of laser-arc hybrid welding to environmental sustainability.
14. Analyze the cost-effectiveness and economic feasibility of laser-arc hybrid welding.
15. Research future trends and potential innovations in laser-arc hybrid welding.

References

1. Seyffarth P, Krivtsun IV (2002) Laser-arc processes and their applications in welding and material treatment. Taylor & Francis, London
2. Cui H, Decker I, Ruge J (1990) Wechselwirkungen zwischen WIG-schweißlichtbogen und fokussiertem laserstrahl. In Waidelich W (ed) Laser/Optoelektronik in der Technik (Laser/optoelectronics in engineering). Springer, Berlin, Heidelberg, pp 577–581
3. Petring D, Fuhrmann C (2004) Recent progress and innovative solutions for laser-arc hybrid welding. In: Pacific international conference on applications of lasers and optics. Laser Institute of America, Melbourne, p 702
4. Gao M, Zeng XY, Hu QW (2006) Effects of welding parameters on melting energy of CO_2 laser–GMA hybrid welding. Sci Technol Weld Joining 11(5):517–522
5. Nilsson K, Engström H, Heimbs S et al (2003) Parameter influence in CO_2-laser/MIG hybrid welding. In: Bucharest: 56th annual assembly of international institute of welding
6. Li RF, Li ZG, Zhu YY et al (2011) A comparative study of laser beam welding and laser–MIG hybrid welding of Ti–Al–Zr–Fe titanium alloy. Mater Sci Eng A 528(3):1138–1142
7. Bidi L, Le Masson P, Cicala E et al (2017) Experimental design method to the weld bead geometry optimization for hybrid laser-MAG welding in a narrow chamfer configuration. Opt Laser Technol 89:114–125
8. Dilthey U, Wieschemann A (2000) Prospects by combining and coupling laser bean and arc welding processes. Weld World 44(3):37–46
9. Chen MH, Xu JN, Xin LJ et al (2017) Effect of keyhole characteristics on porosity formation during pulsed laser-GTA hybrid welding of AZ31B magnesium alloy. Opt Lasers Eng 93:139–145
10. Liu LM, Chen MH, Li CB (2013) Effect of electric arc on laser keyhole behavior based on direct observation during low power pulsed laser–arc hybrid welding process. Opt Lasers Eng 51(10):1153–1160
11. Bagger C, Olsen FO (2005) Review of laser hybrid welding. J Laser Appl 17(1):2–14
12. Cary HB, Helzer SC (2004) Modern welding technology, 6th edn. Pearson, New York
13. Yan J, Gao M, Zeng XY (2010) Study on microstructure and mechanical properties of 304 stainless steel joints by TIG, laser and laser-TIG hybrid welding. Opt Lasers Eng 48(4):512–517
14. Berdnikova O, Pozniakov V, Bushma O (2016) Laser and hybrid laser-arc welding of high strength steel N-A-XTRA-70. Mater Sci Forum 870:630–635

15. Emmelmann C, Kirchhoff M, Petri N (2011) Development of plasma-laser-hybrid welding process. Phys Procedia 12:194–200
16. Kim CH, Ahn YN, Kim JH (2011) CO_2 laser-micro plasma arc hybrid welding for galvanized steel sheets. Trans Nonferrous Metals Soc China 21:s47–s53
17. Zhou J, Zhang T, Tsai H et al (2021) Hybrid laser-arc welding in aerospace engineering. In: Chaturvedi MC (ed) Welding and joining of aerospace materials, 2nd edn. Woodhead Publishing, London, pp 123–156
18. Majumdar JD, Manna I (2012) Introduction to laser assisted fabrication of materials. In: Majumdar JD, Manna I (eds) Laser-assisted fabrication of materials. Springer, Berlin, Heidelberg, pp 1–67
19. Gui H, Zhang K, Li DY et al (2018) Effect of relative position in low-power pulsed-laser–tungsten-inert-gas hybrid welding on laser-arc interaction. J Manuf Process 36:426–433
20. Naito Y, Mizutani M, Katayama S (2006) Penetration characteristics in YAG laser and TIG arc hybrid welding, and arc and plasma/plume behavior during welding-welding phenomena in hybrid welding using YAG laser and TIG arc (first report). Q J Jpn Weld Soc 24(1):32–38
21. Mahrle A, Rose S, Lohse M et al (2014) Interaction mechanisms in hybrid laser arc welding. In: International congress on applications of lasers & electro-optics. Laser Institute of America, San Diego, pp 662–671
22. Chongbunwatana K (2014) Simulation of vapour keyhole and weld pool dynamics during laser beam welding. Prod Eng Res Devel 8(4):499–511
23. Liu LM, Hao XF, Song G (2006) A new laser-arc hybrid welding technique based on energy conservation. Mater Trans 47(6):1611–1614
24. Yan TY, Zhan XH, Gao QY et al (2021) Influence of laser power on molten pool flow field of laser-MIG hybrid welded Invar alloy. Opt Laser Technol 133:106539
25. Zhang W, Hua XM, Liao W et al (2014) The effect of the welding direction on the plasma and metal transfer behavior of CO_2 laser+GMAW-P hybrid welding processes. Opt Lasers Eng 58:102–108
26. Acherjee B (2018) Hybrid laser arc welding: state-of-art review. Opt Laser Technol 99:60–71
27. Leo P, Renna G, Casalino G et al (2015) Effect of power distribution on the weld quality during hybrid laser welding of an Al–Mg alloy. Opt Laser Technol 73:118–126
28. Tani G, Campana G, Fortunato A et al (2007) The influence of shielding gas in hybrid LASER–MIG welding. Appl Surf Sci 253(19):8050–8053
29. Katayama S, Uchiumi S, Mizutani M et al (2007) Penetration and porosity prevention mechanism in YAG laser-MIG hybrid welding. Weld Int 21(1):25–31
30. Han XH, Yang ZB, Ma Y et al (2020) Porosity distribution and mechanical response of laser-MIG hybrid butt welded 6082–T6 aluminum alloy joint. Opt Laser Technol 132:106511
31. Gook S, Midik A, Biegler M et al (2022) Joining 30 mm thick shipbuilding steel plates EH36 using a process combination of hybrid laser arc welding and submerged arc welding. J Manuf Mater Process 6(4):84
32. Górka J, Stano S (2018) Microstructure and properties of hybrid laser arc welded joints (laser beam-MAG) in thermo-mechanical control processed S700MC steel. Metals 8(2):132
33. Wang HY, Liu XH, Liu LM (2020) Research on laser-TIG hybrid welding of 6061–T6 aluminum alloys joint and post heat treatment. Metals 10(1):130

Chapter 3
Electron Beam Welding

3.1 The Basic Principles of Electron Beam Welding

3.1.1 Development of Electron Beam Welding

Electron beam welding has been known for over 60 years. In the last decade, there has been growing interest on the part of individuals in science and industry. Some earlier dates are included, commencing with the first recorded demonstration in 1879 by Sir. Crookes that electrons beam could melt metal. Sir Crookes fused the platinum anode in a cathode ray tube, but it was nearly twenty years later when, in 1897, J. J. Thompson showed that cathode rays were in fact a stream of electrons. The first serious attempt to use the electron beam as a tool for melting appears to be that recorded by Pirani in 1907 [1, 2]. The first electron beam drilling experiment of an unintended hole of Ø 7 mm in a platinum aperture, as well as the first electron beam welding actually bead on plate by K. H. Steigerwald was carried out in 1949. A concentrated beam of electrons was first used for welding in a vacuum chamber in the 1950s in Germany and France. The first reference and photograph of a weld section with a depth to width ratio of the order of 20:1 appears in an anonymous article in Metalworking Production of 20 November 1959, i.e. two years after Stohr's disclosure of the process. As mentioned earlier, the first commercial equipment in the UK was produced towards the end of 1958. All early units available were essentially low-voltage equipment following the original system disclosed by Stohr. Equally, most of the early units were installed at establishments concerned with nuclear energy applications, although they were occasionally used to examine the electron beam weldability of more common materials to ascertain the potential of the technique.

R. Li and T. Li, *Advanced Welding Methods and Equipment*,
https://doi.org/10.1007/978-981-97-4109-0_3

3.1.2 *Working Procedure of Electron Beam Welding*

The working procedure of electron beam welding is shown in Fig. 3.1. After the high voltage current is applied across the cathode tube, the filament present in it heats up and starts producing electrons. The heat is required to maintain at approximately 2500 °C for the constant flow of electrons. These emitted electrons have very little energy. Then, the anode, a positively charged electrode, creates a potential difference between the two terminals. Due to this potential difference, the electrons are accelerated towards the anode.

Further, the highly accelerated electrons pass through the anode and enter into the region having less electric field. Next, it leads to the deflection of some electrons from their path. To focus at one place, the focusing coil is used. The coil focuses the electrons and provides them a focusing trajectory (path).

Then the deflecting coil is used to deflect the electrons on the welding area for higher penetration power. When the highly accelerated electrons strike the surface of metal or workpiece placed on a CNC (Computer Numeric Controller) slab, the kinetic energy of these electrons gets converted into thermal energy. Consequently, it heats up the metal to a molten state. Then by using a CNC machine, a welder adjusts the workpiece for welding. And it cools down to form a weld.

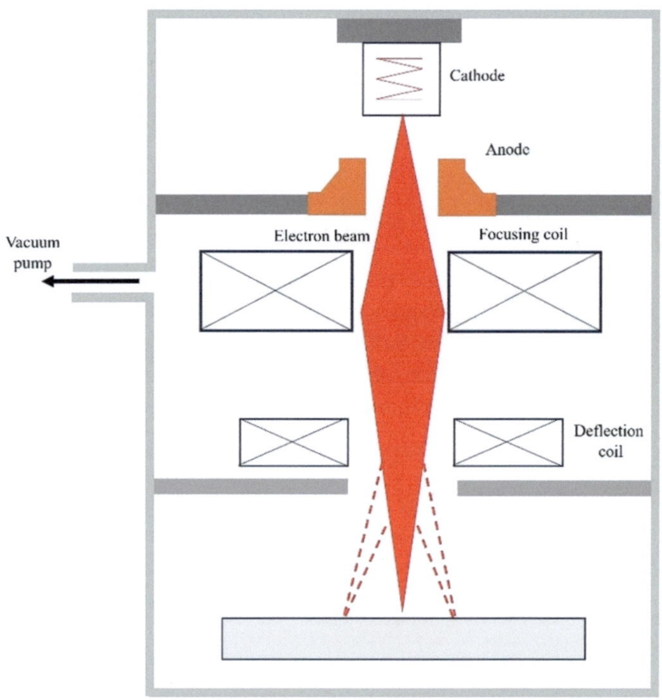

Fig. 3.1 Basic principles of electron beam welding

3.1.3 Characteristics of Electron Beam Welding

In comparison with other methods of joining, electron beam welding has the following characteristics.

(1) Electron beam welding can achieve an extremely high power density of about 10^7 W/cm^2 at the focus of the beam, which is usually higher than that of the laser beam welding.
(2) Beam powers of far less than 1 kW up to 300 kW are available for welding materials with thicknesses of less than 0.5 mm up to 300 mm. The high welding speed can result in narrow welds and heat affected zones with little distortion of the workpiece.
(3) As welding is carried out under vacuum, no consumables (gases, fluxes) are required to protect the weld pool from oxidation. Structural steels, alloy steels, nonferrous metals and even gas-sensitive special metals can be successfully welded.
(4) Short evacuating times can be achieved by adopting the working chamber to suit the number and size of workpieces. Machines are available for welding variable one-off components as well as for use in mass production operations as in the automotive industry.
(5) Computer monitoring and control of the electrical and mechanical welding parameters is possible. The welding parameters, and thus the quality of the welds produced, are highly reproducible and consistent.
(6) The variable working distance allows workpieces of widely differing shapes to be welded. Simple longitudinal weld seams can be welded as well as complicated three dimensional components requiring use of programmed welding parameters and workpiece manipulation.

An electron beam as a heat source has been utilised for many years. However, its use in various technological processes, particularly in welding, is still dynamically developing both in terms of research and industrial applications. The electron beam is also a useful energy source in surface modification, cladding, surfacing, rapid prototyping, engraving as well as texturing [3]. In spite of quickly developing competitive laser welding technologies, electron beam welding remains indispensable in many applications due to the possibility of obtaining greater penetration depth and the metallurgical purity of a weld, as well as because of its greater welding rate. The discussion about the advantages and disadvantages of both the beam welding processes (i.e. electron and laser beam welding) is frequently held in an emotional form. Supporters of both technologies find various advantages and disadvantages of the respective processes. However, the main drawbacks of electron beam welding in vacuum are as follows:

(1) The vacuum limitation. The dimensions of the vacuum chamber imposes an upper limit on the size of component to be handled, since the practice is to place the whole component in the vacuum chamber.

(2) Magnetic materials. Electron beams are readily deflected by small magnetic fields. Thus welding of materials with retained magnetism presents special problems since the alignment of the beam is likely to be distributed by the component's own magnetic field.
(3) Profile 3D welding. The majority of electron beam welding machines are provided with manipulators (x–y table, rotary tilt etc.) capable of producing linear, circular as well as 2D profile welds. Such welds represent by far the largest proportion of all welds produced in industry, but there will be occasions when the design demands a 3D profile welds.
(4) Accessibility. Areas to be reached by the electron beam must be "seen" by the gun; in other words, the path of the electrons from the gun to the area to be welded is a straight line;
(5) Criticality of machining and assembly. The heat produced by a well-focused beam is extremely localized; this is the main asset of the process. It is, however, not without its penalties. Abutment faces have to be machined to close tolerance since a gap above 0.1 mm cannot be tolerated.
(6) Cleaning of the welded elements. Vacuum technologies are quick to point out the importance of cleanliness in association with vacuum plant. The introduction of impurities into the vacuum system can do much to impair the pumping speed and general efficiency of the pumping unit. The other issue is the quality of the welded joints.
(7) Economic consideration. Many factors contribute to the cost of producing electron beam welds such as: the machine hour rate, cost factors, productive time, welding time, etc.

3.2 Electron Beam Welding Equipment

3.2.1 Components of Electron Beam Welding Machine

(1) **Anode**

The anode is a positively charged electrode and used to create the region of an electric field. As it is positively charged, it attracts the electrons towards itself. Then, it leads to the acceleration of the electron.

(2) **Electron gun**

An electron gun or cathode tube ejects the electrons when the temperature of filament presented in it reaches approximately 2500 °C. It is mainly made up of tungsten or tantalum alloy.

(3) **Power supply**

Electron beam welding requires a high voltage power supply. A controller is connected to the power supply so as to control the voltage of current supplied to the electron gun or cathode.

(4) **Focusing coil**

Right after the anode, a focusing coil or convergence coil is present. It mainly acts to focus the electrons on a single path.

(5) **Deflecting coil**

The coil is mainly employed to deflect the focused electron beam on the workpiece.

(6) **Vacuum chamber**

A vacuum chamber is a very important part of the whole process and used to forbid the collision of electrons with the air particles.

(7) **CNC slab**

As the whole process takes place in the vacuum, the workpiece is placed on the Computer Numeric Controller slab. Further, the CNC slab eases the mobility of the workpiece as per requirement.

3.2.2 Classification of Electron Beam Welding Equipment

Electron beam welding processes are usually carried out in a vacuum, yet there are also non-vacuum welding machines available. Modern electron beam welding machines are controlled by PLCs equipped with working tables or numerically controlled welding positioners enabling the automation of welding processes and provided with various control and safety systems aimed to maximize the operator's protection against radiation and carry out technological processes in the vacuum. The division of electron beam welding machines is shown in Fig. 3.2. Types of electron beam welding machines due to working chamber design are shown in Fig. 3.3. Due to technological, production-related and metallurgical requirements, electron beam welding solutions include the following machines [4]:

(1) Universal high pressure welding machines, with a gun fixed inside or outside the working chamber. After appropriate tooling modification, such devices enable welding a vast range of products and elements;
(2) Special high pressure welding machines, intended for welding specific elements such as, for instance, band saws, toothed wheels, turbo-compressor rotors;
(3) Reduced pressure electron beam welding machines with local chambers fixed on a structure being welded. The electron gun is located outside a local chamber of small volume, covering only a section of a flat or girth joint being welded, e.g. of storage tanks. The vacuum is only maintained in the small chamber. Such a solution minimizes the time needed to generate the vacuum;
(4) Non-vacuum electron beam welding machines. The electron beam is generated in high vacuum and at a high accelerating voltage of 150–220 kV. The beam is moved towards the workpiece by the system of vacuum passes, i.e. the system of nozzles gradually reduces vacuum to atmospheric pressure. Welding

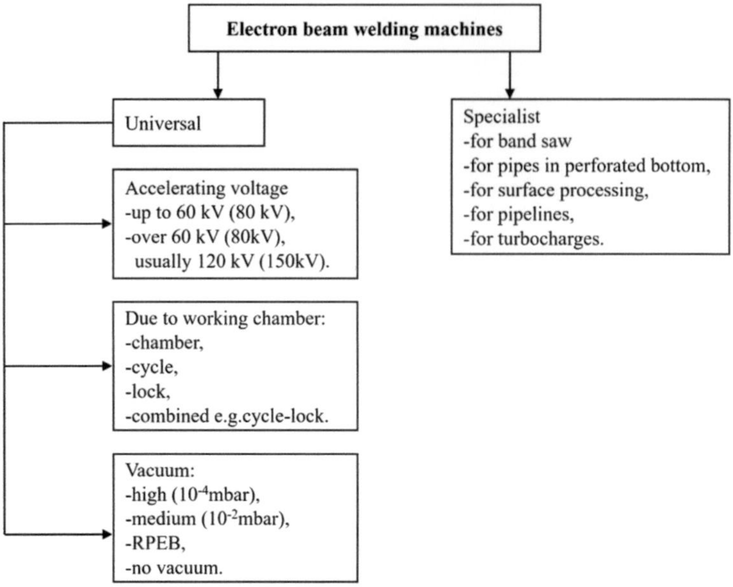

Fig. 3.2 Division of electron beam welding machines

at atmospheric pressure almost entirely eliminates problems related to the size of a structure being welded. The non-vacuum electron beam welding machines are provided with high efficiency pumps and special electron beam discharge orifices to ensure the highest vacuum decrease gradient between the electron gun and atmosphere.

3.3 Process of Electron Beam Welding

3.3.1 Welding Parameters for Electron Beam Welding

The main parameters of the electron beam welding process are the electron beam current I(mA), the accelerating voltage U(V), the welding speed v_{weld} (m/h), the current strength of the magnetic focusing lens I_f(mA), the position of the focus of the beam relative to the surface(mm), the diameter of the heating spot on the surface of the product(mm), the vacuum degree (Pa) and the working distance of the electronic gun from the workpiece(mm).

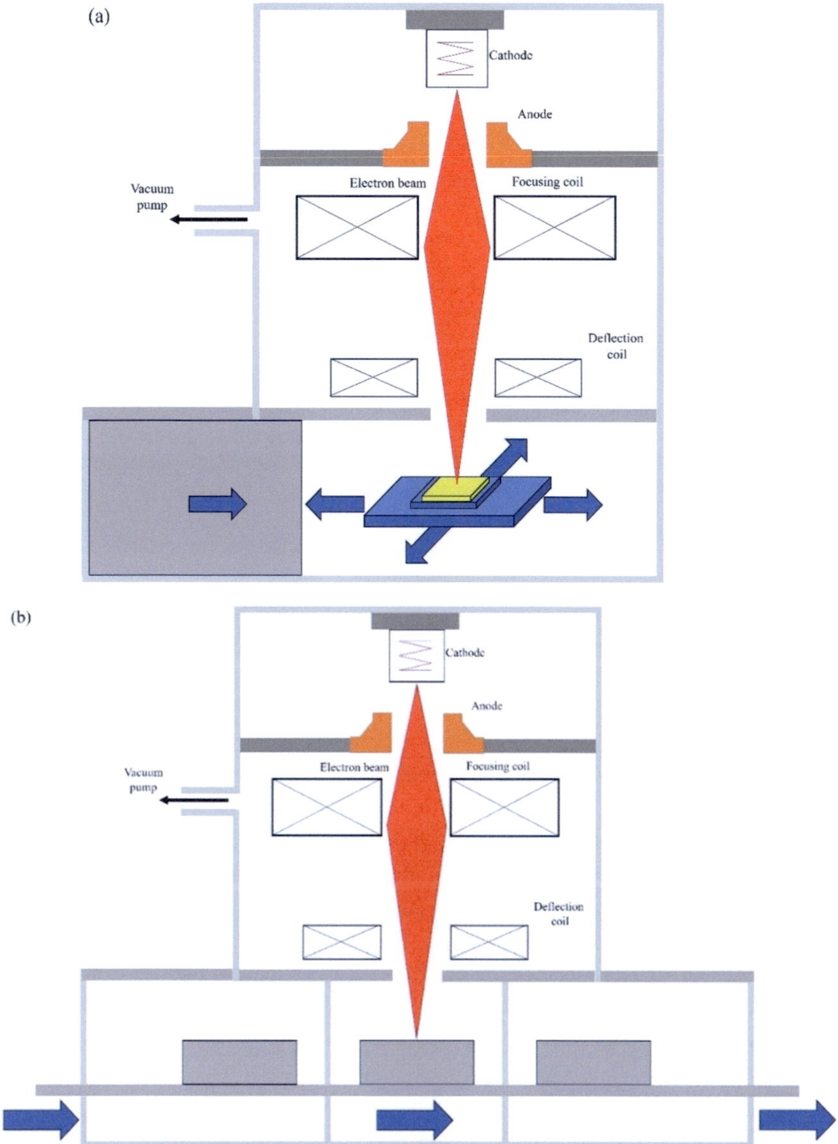

Fig. 3.3 Types of electron beam welding machines due to working chamber design **a** universal welding machine with chamber, **b** cycle machine, **c** continuous operation welding machine (lock)

(c)

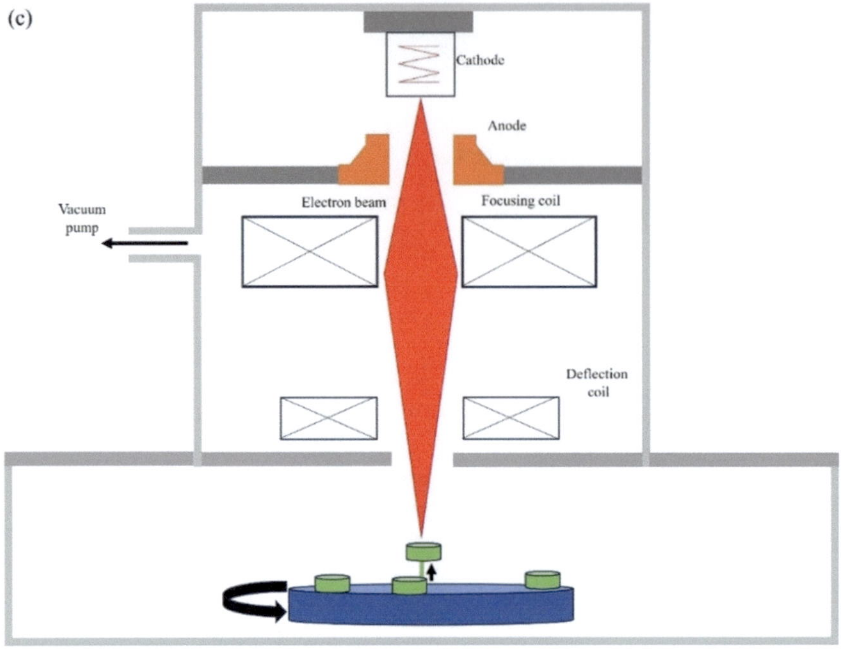

Fig. 3.3 (continued)

3.3.2 *Process of Deep Fusion Welding*

All the processes and installations related to electron beam technology are characterized by the utilization of powerful electron beam generators with beam power up to maxium 100 kW. An electron beam emitted from a tungsten cathode, the most common directly heated cathode, is accelerated by a high voltage in the range of 10–200 kV to over 100 km/s. In the electron beam generator, the electrons emitted by the cathode are bundled into a beam by accelerating and focusing coil. This task is performed by a beam generator which is composed of the cathode, the control electrode and the anode diaphragm located at the earth potential. The beam emerges through the anode diaphragm and spreads in a conical shape until it enters the magnetic field of the focusing lens. After the focusing by the system of focusing lens, the electrons flow through the system of deflecting coils with which the electrons are deflected electromagnetically in a quasi-inertia-free form and can thus be guided over the workpiece [4] (Fig. 3.4). Strongly focused electron beam causes the conversion of the kinetic energy of the electrons into the heat, which is absorbed and melts the welded joint area structure at the energy density of up to 10^{12} W/m^2 [4]. The interaction of the electron beam with the surface of the material is shown in Fig. 3.5. The electron beam strikes the workpiece surface, interacts with the electrons of the material and causes emissions of back scattered electron (reflected), secondary electrons, Auger electrons and X-rays [5].

Fig. 3.4 Schematic of electron beam machine for welding and surface modification

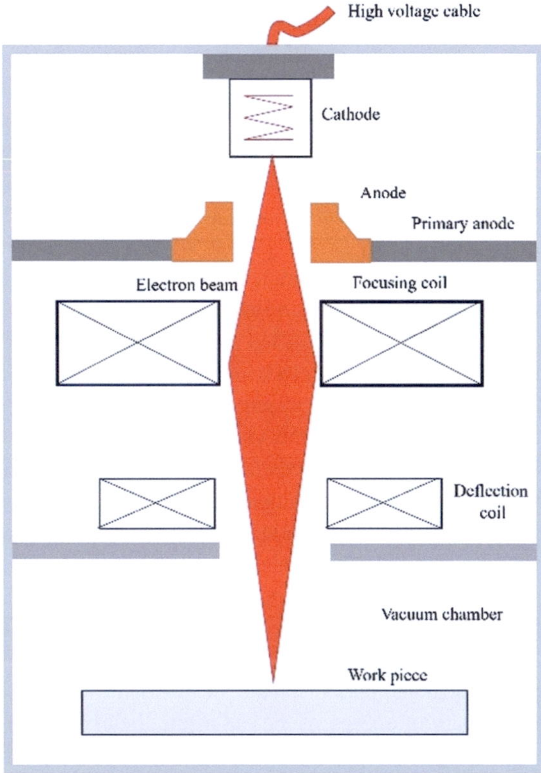

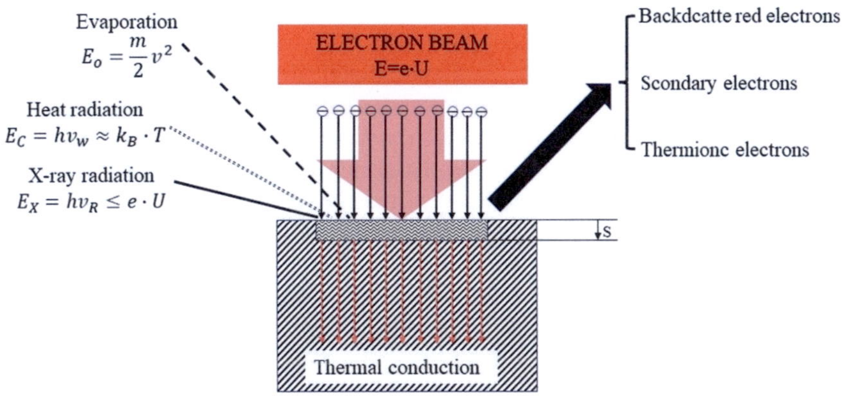

Fig. 3.5 Schematic of interaction between electron beam and surface of the welded material

Each types of radiation are emitted from different depths of the material, always with increasing energy of the primary electron beam (i.e., the increase of the accelerating voltage). Additionally, the depth of penetration of electrons is affected by the atomic number of the material of the welded components. The Auger electrons are emitted from a few atoms, i.e. about 1 nm, thick subsurface layer. These secondary electrons have a characteristic energy depending on the type of scattering atoms. After the electron beam power density reaches the level of 10^{10} W/m^2, the amount of supplied energy is so high that the heat abstraction is not possible by the thermal conductivity of the material, convection or heat radiation. The welded metal at the point of the electron beam impact not only melts but begins to evaporate. A narrow channel is then formed, filled and surrounded by a vapor of molten metal envelope, through which the electron beam penetrates into the deeper layers of the material [6]. The balance of forces retaining channel in equilibrium is shown in Fig. 3.6. The capillary (gasodynamic channel) is formed mainly due to impacts of the metallic vapor pressures and recoil forces. Another effect of electron interaction with material is the emission of X-rays. A continuous spectrum is produced by breaking the incident electrons in the electrostatic field of atomic nuclei. In contrast, the characteristic X-ray emission arises as a result of energy emission through an electron jumping from a higher orbit to the free space on the lower orbit, taking place of the electron being knocked out by the electron of the primary beam. The wavelength and energy of the radiation depend on the atomic number of the element and are not dependent on the physical and chemical state of the material. Based on the analysis of wavelength or intensity of energy of the radiation, the chemical composition of the material (in electron microscopes) can be quantitatively and qualitatively determined [5]. Moreover, the primary electrons excite oscillations of the crystal lattice [7]. An electron beam with an accelerating voltage of 30 kV and current of 150 mA has a big impact on non-alloy steel. Electromagnetic force with the most commonly used ranges of beam current is negligible [8]. The envelope of liquid metal surrounding the capillary has an uneven thickness. It is very thin from the side of the melting zone and wider in the area of the weld solidification [8]. Thermocapillary force occurs because of the differences in surface tension forces caused by the temperature gradient between the melting zone and solidification zone. Transport of liquid metal in the area of the passage takes place around the bundle (in the plane perpendicular to the electron beam) in a direction opposite to the welding direction and in the direction of the electron beam and the counterclockwise [6]. The thermo-capillary convection due to the temperature gradient is known as the Marangoni effect (also the Gibbs-Marangoni effect) [9, 10].

At the bottom of the capillary, the electron beam acting on the welded materials loses. The action is mainly collisions with the ionized metal vapor, molten metal splashes. This causes a periodic change of vapor pressure in the channel and the channel can be closed under the force of surface tension by a thin (several micrometers) film of liquid metal. The X-ray results showed that the capillary can also be closedby the liquid metal pushed out of the bottom of capillary. Closing takes place then in the place where the kinetic energy of the displaced fluid flow decreases to zero. Further, the continuous interaction of the electron beam in turn causes the

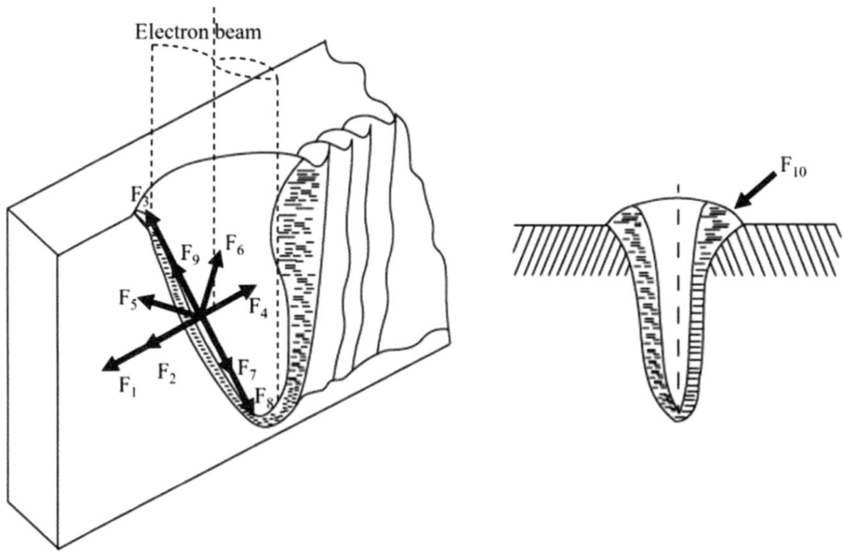

Fig. 3.6 Forces which maintain the cavity in equilibrium: F1: force due to evaporated metal atoms, F2: force due to vapor pressure in cavity, F3: friction between stream of vapor and liquid film, F4: surface tension, F5: electromagnetic force, F6: thermocapillary force due to difference between temperature of front and back walls of cavity, F7: force due to electron pressure, F8: force due to hydrostatic pressure, F9: thermocapillary force due to temperature gradient of front wall, F10: surface tension

melting of metal layers. This process can be done periodically with a frequency of 150 Hz. This point under certain conditions can move at speeds of up to 6.5 m/s. These oscillations occur both at the point of stationary impact of a beam on material, and during welding with full and incomplete penetration [6].

3.4 Applications for Electron Beam Welding

3.4.1 Electron Beam Welding of Large Thick Parts

Electron beam welding is widely used in large thick parts, such as ships, bridges, pressure vessels, etc., which can achieve high-quality and high-efficiency welding, and improve structural strength and durability [11].

Large and thick parts usually bear huge loads and pressures, and welding is a key link to connect structures and transmit strength. Through effective welding of large and thick parts, the integrity and reliability of the structure can be ensured, thereby improving the safety of engineering projects. In addition, the welding of large and thick parts has a greater impact on production efficiency and cost in the engineering field. Large and thick parts usually require a lot of material and human resources, and efficient welding methods can speed up the manufacturing process and reduce production costs. The application of automation and intelligent technology in the welding of large and thick parts can improve the welding speed and quality, and further promote the progress of engineering projects. In short, the welding of large and thick parts is very common and important in the engineering field. It provides key connections and guarantees for various engineering projects, and has an important impact on the safety, efficiency, and cost of the project.

However, the welding of large and thick parts requires high heat input, which leads to an increase in the temperature of the welding area, and thus is prone to residual stress and deformation, affecting the stability and performance of the structure. In addition, more complicated seam preparation work is required during the welding of large and thick parts, including preheating, pretreatment, groove preparation, etc. These processes have high requirements on equipment and technology, which increases the complexity and difficulty of welding. Moreover, it is difficult to monitor welding quality in welding of large and thick parts, defects such as pores and cracks inside the weld are often difficult to detect, and complex non-destructive testing techniques are required for quality control.

Therefore, in order to overcome these difficulties, it is usually necessary to use electron beam welding technology and equipment to improve the efficiency and quality of welding large and thick parts, and to ensure the stability and reliability of the structure.

Electron beam welding is essential in thick plate welding. An example of electron beam welding of a 50 mm thick plate is shown in Fig. 3.7. This advanced welding method excels in joining heavy-duty components, such as those found in shipbuilding and large-scale construction projects. The image showcases the deep penetration and exceptional fusion quality achievable with EBW. Its high energy density ensures robust and high-strength welds in massive structures. Reduced heat input minimizes distortion and residual stress, critical in thick plate welding where dimensional accuracy and structural integrity are paramount. Electron beam welding is indispensable for industries requiring reliable, durable, and defect-free welds in substantial and demanding applications.

3.4.2 Applications of Electron Beam Welding in Aerospace

Electron beam welding is widely used in the aerospace industry for joining complex metal structures such as aero-engine parts, airframe structures, etc. [13, 14]. The welded joints of the spacecraft need to have high strength, high sealing, and high reliability. In aircraft manufacturing, electron beam welding finds application in joining structural elements such as fuselage sections and wing assemblies. Its capability to provide deep penetration and narrow heat-affected zones contributes to high-quality welds with minimal distortion. The resulting welds possess exceptional strength, contributing to the overall structural integrity and safety of the aircraft.

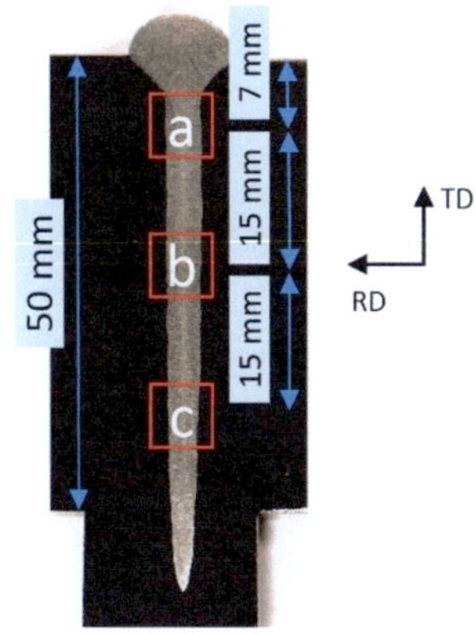

Fig. 3.7 Example of electron beam welding of 50 mm thick plate [12]

Moreover, electron beam welding is extensively employed in the fabrication of spacecraft components and satellite structures. It allows for precision welding of lightweight materials like aluminum and titanium, enabling the construction of robust and lightweight space vehicles. This weight reduction is crucial for achieving optimal performance and reducing launch costs.

Furthermore, electron beam welding supports the manufacturing of propulsion systems and rocket engines. It ensures the reliable joining of complex geometries and dissimilar materials, facilitating the production of efficient and high-performance rocket engines. The high energy density of the electron beam enables deep penetration welding of thick sections, enhancing the structural integrity of these critical components.

In summary, electron beam welding plays a vital role in the aerospace industry. Its ability to deliver high-quality welds with minimized distortion, superior strength, and excellent fatigue resistance makes it indispensable for various applications such as engine components, aircraft structures, spacecraft, and rocket engines. Electron beam welding contributes to the advancement and success of aerospace technology by ensuring the highest standards of quality and safety.

Figure 3.8 shows the microscopic morphology of a Ti6Al4V electron beam welded joint. Ti6Al4V's high strength-to-weight ratio and corrosion resistance make it a preferred material for aerospace structures. EBW's capability to weld this challenging alloy with minimal heat input safeguards its mechanical properties. The resulting welds exhibit outstanding structural integrity and performance, meeting the stringent safety and reliability requirements of the aerospace industry.

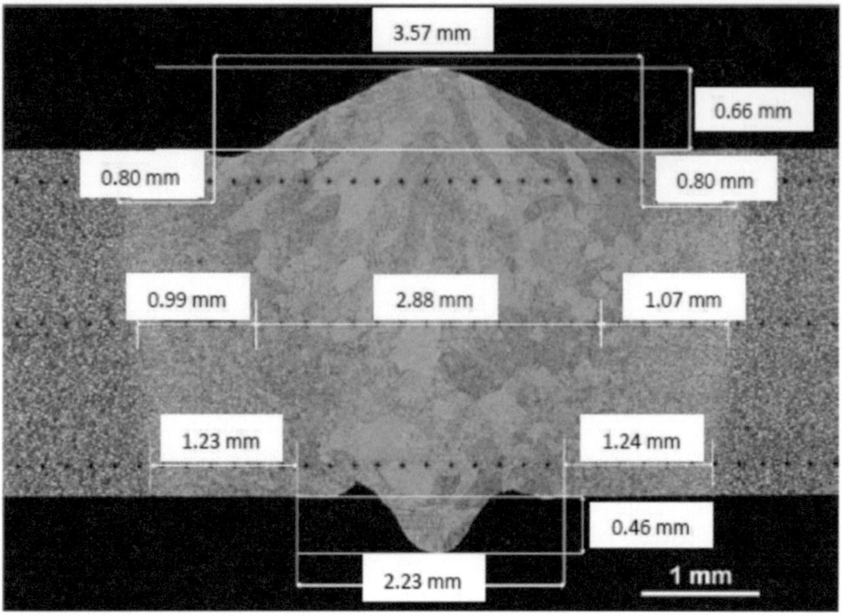

Fig. 3.8 Shape of the weld together with the size of the heat-affected zone [15]

3.4.3 Applications in the Electronics and Instrumentation Industry

Electron beam welding is widely used in electronic device manufacturing to connect electronic components, circuit boards, and microscopic electronic devices. It can provide precise welding control and tiny solder joints.

Welding is a common and important in the electronic instrumentation industry, which often contains complex circuit boards and components that need to be connected and fixed. Welding methods are one of the most commonly used joining techniques to create electrical and mechanical connections between electronic components [16, 17].

Traditional welding methods have some difficulties and limitations in the electronic instrument and meter industry. The first is that traditional welding methods have a greater impact on fine parts. Electronic instruments and meters usually contain tiny and complex parts, such as IC chips, miniature resistors, and capacitors. The impact of traditional welding methods on these delicate parts is a challenge. High temperatures and thermal stress during welding can cause thermal damage, deformation, or failure of parts [18].

The second is the influence of heat input and heat-affected regions on circuit electronics, where heat input is a major problem. Welding introduces heat that can cause nearby electronic components to be subjected to thermal stress or thermal expansion. This can lead to failure of circuit elements or reliability issues. At the same time, diffusion in the heat-affected region causes changes in material properties, such as resistivity, electrical conductivity, etc., which in turn affect the performance and accuracy of electronic instruments and meters [19].

In summary, traditional welding methods have some difficulties and limitations in the electronic instrument and meter industry. Electron beam welding has high energy density, precise control, and minimal heat-affected zone. It is beneficial for the application in electronic instrument manufacturing.

One prominent application of electron beam welding is in the assembly of electronic enclosures and housings. It allows for the precise welding of complex and delicate components, ensuring a hermetic seal for sensitive electronic circuitry. The high energy concentration of the electron beam enables efficient welding of thin-walled structures while minimizing heat input, reducing the risk of component damage or distortion.

Electron beam welding also finds use in the fabrication of electronic connectors and terminals. It enables the reliable joining of dissimilar materials, such as copper, aluminum, and stainless steel, ensuring excellent electrical conductivity and mechanical strength. The narrow and deep welds achieved through electron beam welding enhance the long-term reliability and performance of connectors, meeting the strict quality standards of the electronic instrument industry.

Additionally, electron beam welding is employed in the production of electronic sensors and transducers. It enables the precise welding of miniature components with minimal heat distortion, ensuring the integrity and accuracy of sensor measurements. The high-quality welded joints achieved through electron beam welding contribute to the durability and reliability of these critical components.

Moreover, electron beam welding supports the manufacturing of electronic display panels and screens. Its ability to provide deep penetration and narrow fusion zones allows for the precise welding of delicate components, enabling the production of seamless and defect-free displays. The minimal heat input ensures that the electronic components remain undamaged during the welding process, ensuring optimal display performance.

Furthermore, electron beam welding plays a role in the production of electronic instrument components such as electron guns and cathodes. Its high energy density and precise control enable the fabrication of intricate and high-performance components, ensuring optimal functionality and longevity.

Figure 3.9 shows the macroscopic morphology of a Ti-4Al-5Mo-5V-5Cr-1Nb alloy election beam welding joint. This alloy's high strength-to-weight ratio and resistance to corrosion make it ideal for precision instrument components. EBW, with its precise control and minimal heat input, ensures welds maintain the alloy's mechanical properties. The resulting joints exhibit outstanding strength, corrosion resistance, and precision, meeting the stringent demands of electronic instrumentation where reliability, dimensional accuracy, and material integrity are critical, making EBW a preferred welding method in this industry.

Fig. 3.9 Macroscopic morphology of butt joint: **a** top surface; **b** bottom surface; and **c** cross-section [20]

3.4.4 Application in the Production of Automotive Parts

Electron beam welding is used in the automotive industry to join frames, body parts, and engine components. It can provide high strength and high-quality welds [21].

Automotive parts often consist of multiple metal parts that need to be connected to build complex automotive structures. Welding is a commonly used joining technology that provides a strong, reliable connection and meets the structural strength and durability requirements of automotive parts [22].

One important application of electron beam welding in the automotive industry is the welding of structural components such as chassis and body frames. The high energy concentration of the electron beam allows for efficient welding of thick sections, ensuring robust and durable joints. The deep penetration capability enables strong welds with minimal distortion, contributing to the overall structural integrity and safety of vehicles.

Electron beam welding is also utilized in the assembly of exhaust systems and mufflers. It enables the precise joining of stainless steel and other high-temperature alloys used in these components. The high energy density of the electron beam ensures reliable and leak-free welds, enhancing the performance and longevity of the exhaust system.

Moreover, electron beam welding finds application in the fabrication of fuel tanks and containers. It enables the welding of thin-gauge materials such as aluminum and steel, providing high-quality welds with excellent resistance to corrosion and leakage. The precise control of the electron beam facilitates the welding of intricate shapes and complex designs, meeting the strict safety standards of the automotive industry.

Additionally, electron beam welding supports the manufacturing of automotive engine components, including cylinder heads, blocks, and crankshafts. It allows for the precise welding of dissimilar materials, such as aluminum and cast iron, ensuring reliable and high-strength joints. The deep penetration capability ensures the integrity and longevity of these critical engine components, contributing to optimal performance and efficiency.

Furthermore, electron beam welding plays a role in the production of automotive electrical components such as battery terminals and connectors. Its precise control and minimal heat input ensure the integrity and reliability of electrical connections. The high energy density of the electron beam enables efficient and defect-free welding, meeting the stringent electrical performance requirements of automotive systems.

In summary, electron beam welding is a valuable technology in the automotive industry. Its advantages, including deep penetration capability, precise control, and high-quality welds, contribute to the manufacturing of structural components, exhaust systems, fuel tanks, engine components, and electrical connections. Electron beam welding ensures the reliability, performance, and safety of automobiles, making it an indispensable process in this field.

Electron beam welding (EBW) finds vital application in the automotive industry, as evidenced by the microstructure of an EN AW-6082 T6 aluminum alloy joint, displayed in the accompanying Fig. 3.10. This technique is instrumental in joining aluminum components used in modern vehicles, ensuring both structural integrity and weight savings. The micrograph illustrates the fine grain structure and absence of macroscopic defects. This precision welding method delivers welds with exceptional mechanical properties, crucial for automotive safety and fuel efficiency. The reduced heat input minimizes distortion, preserving the material's strength, making EBW an indispensable technology for achieving lightweight, high-performance automotive structures.

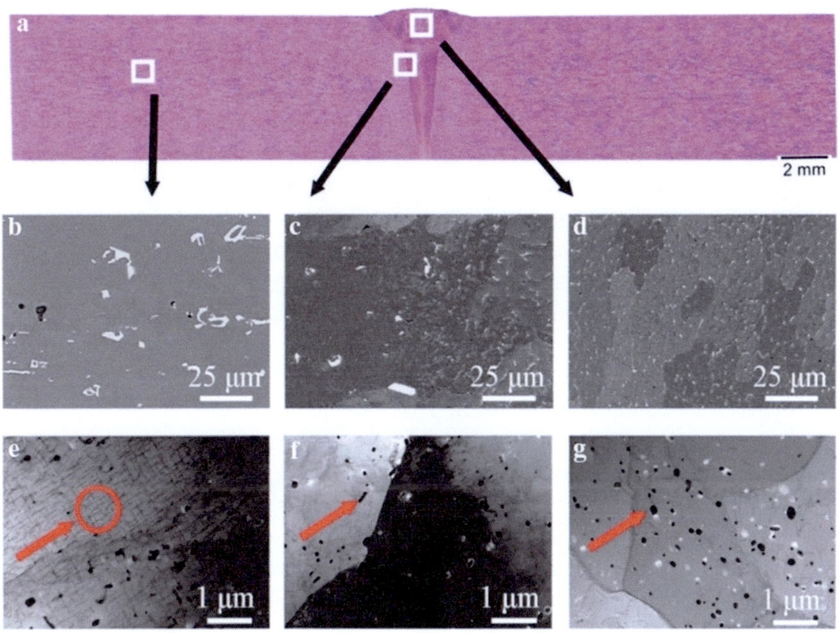

Fig. 3.10 The microstructure of EBW joints: **a** light microscopy; **b** base material (SEM); **c** heat-affected zone (SEM); **d** weld zone (SEM); **e** base material (TEM); **f** heat-affected zone (TEM); **g** weld zone (TEM) [23]

3.5 Other Electron Beam Processing Method

In addition to electron beam welding, the heat source of electron beam can also be used for cutting, surface modification, ectching, additive manufactruing, and so on.

3.5.1 Electron Beam Cutting

1. **Electron beam cutting principle**

Electron beam cutting is a processing method that uses high-energy electron beams to cut materials. Figure 3.11 shows a schematic diagram of the electron beam cutting principle [24]. The focused electron beam is guided to the target material, and the cutting path is controlled by scanning or controlling the pulse. The high energy of the electron beam allows it to penetrate the material and cut it open by stripping, melting, or evaporating. During electron beam cutting, auxiliary gases such as oxygen, nitrogen, or inert gas are usually introduced. Auxiliary gases help remove melt from the cutting area, reduce material splashes and cool cut edges.

2. **Electron beam cutting equipment**

Electron beam cutting equipment consists of different components and comes in different types. The components of electron beam cutting equipment are as follows.

(1) Electronic source: The electron source is the core component of electron beam cutting equipment, and an electron gun is usually used as the electron source. The electron beam is generated and accelerated by pyroelectric or applied voltage in the electron gun [25].
(2) Focusing system: The focusing system is used to focus the electron beam from the electron gun into a bundle and focus it on the cutting point. It usually consists of lenses, magnetic fields, or electric field focusing components.
(3) Cutting table: The cutting table is a platform for placing the material to be cut, with a movable mechanical structure for locating and controlling the position of the material to be cut. The table can also be fitted with a device for lifting the material so that the electron beam passes through the cut.
(4) Control system: The electron beam cutting equipment is equipped with a control system for controlling and monitoring the parameters and cutting process of the electron beam [26]. This includes the electron gun's accelerating voltage, focusing parameters, cutting speed, scanning path, etc.
(5) Auxiliary gas system: Introducing auxiliary gas into electron beam cutting can help remove melt from the cutting area, cool the cutting edge, and improve cut quality. The auxiliary gas supply system consists of a gas source, a regulating device and a nozzle.

According to the way the electron beam is generated and focused, the electron beam cutting equipment can be divided into the following categories.

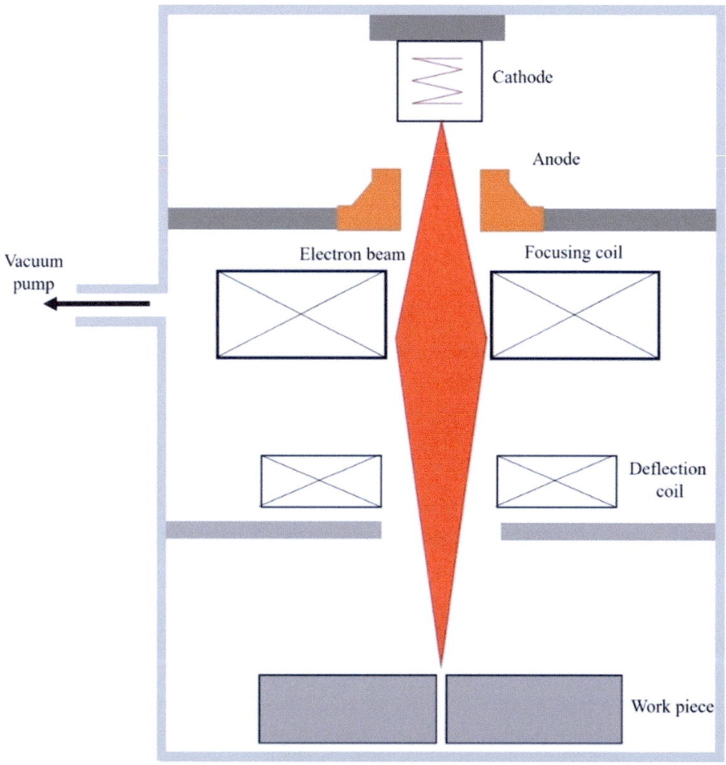

Fig. 3.11 Principle of electron beam cutting

(1) Electrostatic focused electron beam cutting equipment: Electrostatic focused electron beam cutting equipment uses electrostatic fields to focus electron beams, mainly used for cutting metal materials. It has a large cutting depth and high cutting speed.
(2) Magnetic focusing electron beam cutting equipment: Magnetic focusing electron beam cutting equipment focuses the electron beam through the magnetic field, which is widely used in the cutting of metal and non-metal materials. Magnetic focused electron beam cutting equipment has high precision and controllability.
(3) Photoelectron beam cutting equipment: Photoelectron beam cutting equipment uses a strong laser to heat the surface of the material and generate an electron beam, and then uses optical components such as lenses to focus and cut. Photoelectron beam cutting equipment is suitable for cutting various materials, with the advantages of high speed, high precision, and no contact.

(4) Electron beam cutting mechanical system: The electron beam cutting mechanical system is a multi-axis CNC equipment, which realizes cutting in all directions by controlling the movement of the electron beam cutting head. This equipment is suitable for cutting workpieces with large and complex structures.

3. **Process flow of electron beam cutting**

(1) Material preparation: First, you need to prepare the material to be cut. This can be a sheet of metal, a plastic plate, or a ceramic block, etc. The size and thickness of the material affect the settings and parameters of the subsequent cutting process.
(2) Setting cutting parameters: According to the characteristics of the material and the required cutting result, it is necessary to set the cutting parameters. This includes the energy of the electron beam, focus, scanning speed, scanning shape, etc. The setting of cutting parameters directly affects cutting quality and efficiency.
(3) Determining the cutting path: According to the shape and profile of the cut, the cutting path needs to be determined. This allows the desired shape to be drawn by CAD software and converted into machine-readable instructions.
(4) Starting the cutting equipment: Once the cutting parameters and path are set, the electron beam cutting equipment for cutting is started. At this point, the electron beam cutting head moves to the initial position, ready to start the cutting process.
(5) Auxiliary gas injection: Auxiliary gas is usually introduced during the cutting process. The selection of auxiliary gases and injection parameters need to be adjusted according to the characteristics of the material.
(6) Electron beam cutting: Once the equipment is ready and the auxiliary gas injection is normal, start cutting. The electron beam is shot from the cutting head and moves across the material through the scanning path. The high energy of the electron beam causes the material to melt, evaporate, or peel off to achieve cutting.
(7) Cutting control: During the cutting process, the position and speed of the cutting head need to be controlled to ensure the accuracy and consistency of the cutting path. The control system controls the switching and movement of the electron beam according to the preset cutting path
(8) Cutting quality inspection: After the cutting is completed, the cutting quality needs to be checked. Light microscopes or other inspection equipment can be used to check the smoothness, roughness, and accuracy of the cut shape.

4. **Application of electron beam cutting**

Electron beam cutting is widely used in various industrial fields due to its high precision, high energy, and non-contact characteristics. Here are a few of the main application directions for electron beam cutting [24]:

(1) Metal processing: Electron beam cutting is widely used in metal processing. It can be used to cut metal materials such as steel plate, aluminum alloy, stainless steel, etc.

(2) Automotive manufacturing: Electron beam cutting plays an important role in the automobile manufacturing process. It can be used to cut automotive body structure parts, chassis parts and interior decoration parts.
(3) Aerospace: Electron beam cutting has important applications in the aerospace field. It can be used to cut turbine blades in aero engines, aerospace structural parts, and shells and components of avionics.
(4) Electronic device manufacturing: Electron beam cutting also plays an important role in electronic device manufacturing. It can be used to cut small and brittle materials such as integrated circuits, semiconductor devices, optical components, etc.

In summary, electron beam cutting is widely used in metal processing, automobile manufacturing, aerospace and electronic device manufacturing and other fields due to its high precision, high energy, and non-contact characteristics. As technology continues to evolve and innovate, the application of electron beam cutting in many other fields continues to expand.

3.5.2 Electron Beam Surface Modification

1. **Electron beam surface modification principle**

Electron beam surface modification is a method of treating the surface of a material using a high-energy electron beam. Figure 3.12 shows a schematic diagram of the electron beam surface modification principle. Its basic principles can be summarized in the following aspects [27]:

(1) Energy transfer: The interaction of high-energy electron beams with the surface of the material leads to energy transfer. After the electron beam enters the surface of the material, it collides with the atoms of the material, transferring energy to the atoms. This energy transfer causes processes such as displacement, vibration, and ionization of atoms.
(2) Melting and resolidification: The energy density of high-energy electron beams is high, which can melt the surface of the material in a short time. The melted material cools rapidly under the action of an electron beam to form a rapidly solidified microstructure. This rapid solidification process can lead to changes in the crystal structure and phase transitions of the material.
(3) Crystal defect formation: The incidence of high-energy electron beams will cause the displacement and ionization of material atoms, and form crystal defects, such as dislocations, vacancies, and gaps. The formation of crystal defects can change the mechanical properties, thermal conductivity, and electrical conductivity of the material.
(4) Nitriding and carbonization reaction: In the process of electron beam surface modification, in addition to energy transfer and melting solidification, auxiliary gases, such as nitrogen or methane, can also be introduced to undergo nitriding,

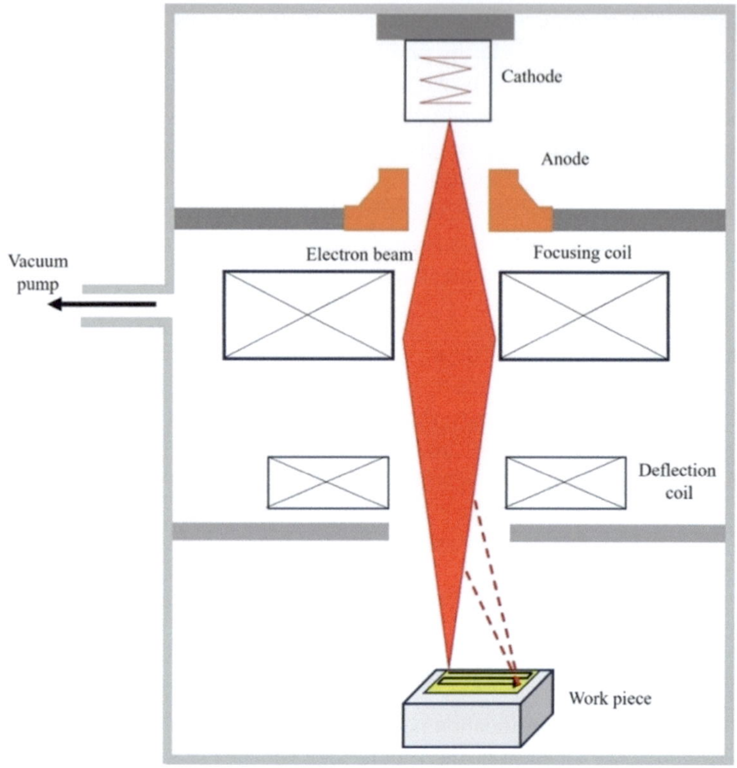

Fig. 3.12 Schematic diagram of electron beam surface modification

carbonization or alloying reactions with materials. These chemical reactions can change properties such as the chemical composition, strength, and wear resistance of a material.

2. **Electron beam surface modification equipment**

The electron beam surface modification equipment is mainly composed of the following components:

(1) Electron beam generator: The electron beam generator is the core component of electron beam surface modification equipment to generate high-energy electron beams [24]. Common electron beam generators include hot cathode electron guns and field-emitting electron guns. Among them, the hot cathode electron gun uses the hot cathode in the electron gun to generate an electron beam, while the field emitting electron gun uses an electric field or magnetic field to accelerate the generation of an electron beam.
(2) Platform or workbench: A platform or workbench is where the material to be processed is placed. It usually has adjustable and positionable functions to ensure accurate positioning of the material with the electron beam. The platform

or table may have functions such as swivel, lift, and tilt to provide a more flexible handling scheme.

(3) Auxiliary gas system: The auxiliary gas system is used to introduce auxiliary gas during electron beam surface modification. Auxiliary gases can facilitate reactions and cooling under the action of high-energy electron beams to change the physical and chemical properties of the material surface. The auxiliary gas system includes a gas source, a control unit, and a nozzle.
(4) Control system: The control system is used to control the parameters of electron beam surface modification equipment [26]. This includes control of aspects such as the energy, density, and scanning speed of the electron beam. The control system can also monitor and record changes in key parameters during electron beam surface modification to ensure consistent processing.
(5) Sensors: Sensors are used to monitor key parameters in the electron beam surface modification process. Common sensors include temperature sensors, pressure sensors, and electron beam intensity sensors. By monitoring these parameters in real time, the electron beam surface modification process can be optimized.

Electron beam surface modification equipment can be divided into the following categories according to different working principles and structural characteristics [24]:

(1) Electron beam impact equipment: Electron beam impact equipment is the most basic electron beam surface modification equipment [23]. Through the high-energy impact of the electron beam, the surface of the material is instantaneously melted and resolidified, thereby changing the crystal structure and properties of the material. Such devices typically consist of an electron beam generator, a beam control system, and a working platform.
(2) Electron beam surface alloying equipment: This type of equipment reacts the alloyed elements with the surface of the material to form new compounds or alloy layers by introducing alloying elements in the process of electron beam surface modification. Such equipment typically includes electron beam generators, auxiliary gas systems, control systems, and alloying sources.
(3) Electron beam sputtering equipment: Electron beam sputtering equipment uses a high-energy electron beam to hit the target material, causing it to evaporate or sputter, and deposit it on the surface of the substrate to form a thin film. This equipment often consists of an electron beam generator, a thin film material source, a substrate fixture, and a deposition chamber. Electron beam sputtering enables the control of the surface composition and structure of the substrate to produce films with specific properties.
(4) Electron beam annealing equipment: Electron beam annealing equipment realizes the annealing treatment of materials through the irradiation of high-energy electron beams. Electron beam annealing can eliminate the internal stress, crystal defects and amorphous structure of the material, and improve the mechanical and thermal stability of the material. Such devices typically include electron beam generators, control systems, and heating devices.

(5) Electron beam atomization equipment: Electron beam atomization equipment is a method for preparing nanoparticles and powders. Through the high-energy action of the electron beam, the surface of the raw material is evaporated and sputtered to form steam and particles, which are then condensed into nanoparticles. Such equipment typically includes electron beam generators, evaporation chambers, coalescing devices, and collection systems.

3. **Process flow of electron beam surface modification**

The basic process steps of electron beam surface modification are described below [28, 29]:

(1) Material preparation: First, choose the appropriate material, determine its shape and size. For materials with complex shapes, machining and adjustment may be required.
(2) Set process parameters: according to the specific surface modification requirements, determine the appropriate process parameters such as electron beam energy, scanning speed, gas environment, etc. These parameters will directly affect the modification effect of the material surface.
(3) Adjust the electron beam: According to the set process parameters, adjust the electron beam generator to generate a high-energy electron beam. The beam's convergence, energy, density, etc. are usually adjusted.
(4) Electron beam treatment: The material is placed on the working platform of the equipment, and the movement and rotation of the platform are controlled so that the electron beam can evenly irradiate the surface of the material. The high energy of the electron beam will produce melting and resolidification in interaction with the surface.
(5) Auxiliary gas treatment: Depending on the needs, auxiliary gases, such as nitrogen or methane, can be introduced to promote chemical reactions during electron beam irradiation. The selection and control of auxiliary gases can change the chemical composition and structure of the material surface
(6) Real-time monitoring: Through sensors and sensors, real-time monitoring of process parameters and key parameters of surface modification, such as temperature, pressure, electron beam irradiation intensity, etc. This monitoring data can be used to adjust process parameters to ensure consistent processing results.
(7) Surface analysis and evaluation: After the electron beam surface modification is completed, the material surface is analyzed and evaluated. Methods such as scanning electron microscopy, X-ray diffraction, and chemical analysis can be used to evaluate changes in the chemical composition, crystal structure, and physical properties of surfaces.
(8) Manage post-processing: According to actual needs, post-treatment may be required, such as annealing, polishing, coating, etc., to further improve the performance and appearance of the material surface.

4. **Application of electron beam surface modification**

Electron beam surface modification technology has a wide range of applications due to its ability to precisely control the physical, chemical, and structural properties of material surfaces. The following are a few of the main applications of electron beam surface modification [30]:

(1) Material strengthening: Metallurgical phase can be formed on the surface of the material through electron beam surface modification, increasing the hardness, wear resistance and corrosion resistance of the material, thereby improving the strength and durability of the material. This is widely used in fields such as aerospace, automotive manufacturing, and mechanical engineering.
(2) Surface alloying: Electron beam surface modification can form alloy layers or compounds on the surface of the material, increasing the wear resistance, corrosion resistance and thermal stability of the material while maintaining the macroscopic properties of the substrate. This has important applications in the fields of chemicals, electronics, and shipbuilding.
(3) Surface coating: Electron beam surface modification can be used to deposit functional films and coatings, such as metals, ceramics, oxides, etc. Such films can provide specific properties on the surface of materials, such as electrical conductivity, thermal insulation, and optical properties, and are widely used in optics, electronics, and energy materials.
(4) Interface modification: electron beam surface modification can improve the interface properties of materials and surrounding environments or other materials, such as increasing interface adhesion strength, reducing interface friction and interface reaction. This plays an important role in areas such as bonding technology, coating engineering and composite material manufacturing.
(5) Preparation of nanoparticles: Nanoparticles and nanostructured materials can be prepared by using electron beam surface modification technology, which is of great significance in nanoscience and nanotechnology research, and has important application prospects for the preparation of nanomaterials and nanodevices [31].

3.5.3 Electron Beam Etching

1. **Electron beam etching principle**

Electron beam etching is a technique that uses high-energy electron beams for surface etching. Figure 3.13 shows a schematic diagram of the electron beam etching. The basic principle can be summarized in the following steps:

(1) Generation of high-energy electron beams: First, a high-energy electron beam is generated using an electron gun or similar device. The electron beam is usually accelerated by an accelerator to reach high energy levels.
(2) Control of gas environment: In the process of electron beam etching, the type and pressure of the gas should be controlled. Typical gases can be oxygen, fluorine,

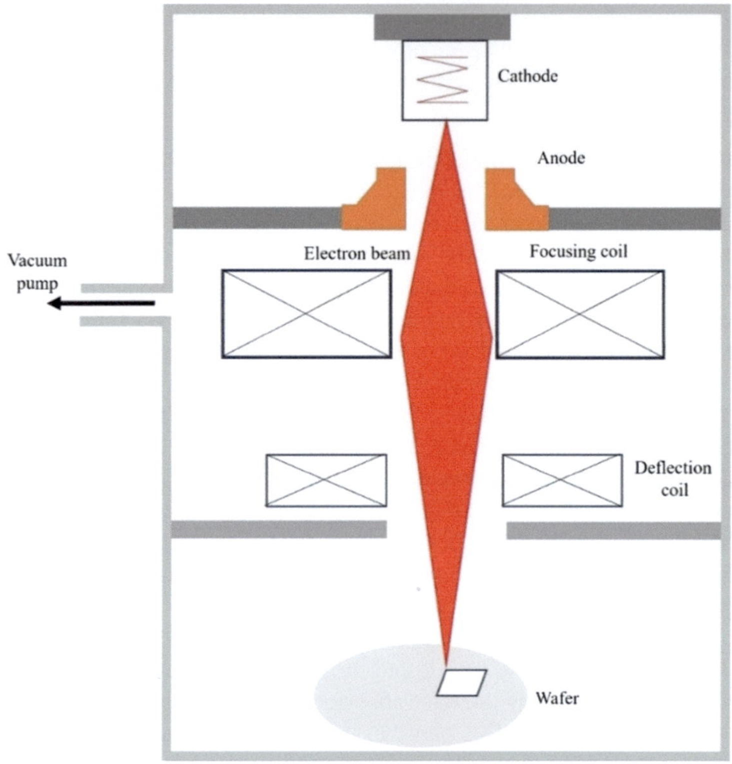

Fig. 3.13 Schematic diagram of electron beam etching

or chlorine. These gases undergo chemical reactions or simple collisions under high-energy electron beams, causing etching of the surface of the material.

(3) Focusing and control of the electron beam: The electron beam needs to go through a series of focusing and control before irradiating the surface of the substance. Usually, magnetic field or electric field is used to control the direction and intensity of the beam to ensure the precise positioning and focus of the beam, so as to achieve the etching of the surface of the material.

(4) Hitting and etching on the surface of the material: After the high-energy electron beam hits the surface of the material, the electrons will interact with the atoms or molecules on the surface. This can lead to ionization of atoms or molecules, breaking of bonds, fragmentation of atoms, etc., so that surface atoms or molecules are removed to achieve an etching effect.

(5) Control and removal of etchants: In the process of electron beam etching, the stability and effectiveness of the etchants should be ensured. In addition, the by-products of the etched material need to be removed by a gas stream or extraction system to ensure the cleanliness of the etched area and the consistent etching quality.

2. **Electron beam etching equipment**

Electron beam etching is a specialized device used to implement the electron beam etching process. The equipment consists of several key components to achieve precise etching of the surface of the material. Electron beam etching equipment is mainly divided into the following components:

(1) Electronic gun: One of the core components of the electron beam etching equipment is the electron gun [32]. The electron gun is responsible for generating high-energy electron beams. It typically includes a cathode and an anode, as well as a high-voltage power supply and focusing elements for accelerating the electrons. The cathode in the electron gun is usually a hot cathode or a field-induced emission (FEG) cathode [33].
(2) Acceleration system: The acceleration system is used to provide the high energy required by the electron beam. It consists of a high-voltage power supply and a series of accelerating electrodes. The acceleration system accelerates the electron beam based on the set energy, allowing it to obtain enough kinetic energy for the etching process.
(3) Gas control system: In the electron beam etching process, it is necessary to select the appropriate gas environment. Gas control systems are used to supply and control the gas in the work area to achieve the desired etch reaction. Commonly used gases can include oxygen, fluorine, or chlorine.
(4) Focusing system: The focusing system of the electron beam etching equipment is responsible for focusing the electron beam to the desired area to achieve precise etch position and resolution. A focusing system usually consists of a set of magnetic field electrodes or electric field lenses that focus the beam by changing the strength and shape of the magnetic field or electric field.
(5) Sample stage or fixture: The sample stage or fixture is used to fix the material sample to be etched to maintain its stable position and posture during the etching process. The sample stage usually has a multi-axis motion function to facilitate precise positioning and rotation of the sample, so as to achieve etching of different areas of the sample.
(6) Control system: Electron beam etching equipment also includes a control system for controlling and monitoring various parameters in the etching process. The control system can include the control and adjustment of parameters such as electron beam energy, beam density, irradiation time, gas flow, and sample movement. It can also be equipped with a real-time monitoring system for monitoring parameters such as surface topography and etch rate during the etching process.
(7) Pumps and pumping systems: In electron beam etching equipment, it is necessary to ensure a vacuum environment to avoid interference with electron beam etching by oxygen or other gases. The pumping system includes vacuum pumps and associated equipment for the removal of gases and products from the etched area.

Electron beam etching equipment can be classified according to different classifications. The following are two common classifications for electron beam etching equipment:

(1) Classification according to etching mode: According to the etching mode, electron beam etching equipment can be divided into two categories: direct etching and indirect etching. Direct etching refers to the electron beam acting directly on the surface of the material, stripping atoms or molecules from the surface to achieve etching. Indirect etching refers to etching by the excitation and dissociation of electron beams in compressed gases to produce active particles.
(2) Classification according to the scale of etching: According to the scale of etching, electron beam etching equipment can be divided into large equipment and small equipment. Large-scale equipment is mainly used for mass production or mass etching, often with high-energy electron beams, large-sized sample stages, and multiple working chambers, suitable for industrial applications such as semiconductor manufacturing and display production. Small equipment is mainly used in research and development fields, usually with smaller size, more flexible operation, and higher resolution, suitable for research laboratories and academic research institutions.

These classifications can help people understand the different characteristics and application scenarios of electron beam etching equipment, so as to choose the right equipment according to specific needs. Electron beam etching is a process that uses high-energy electron beams for surface etching. It can achieve high-precision and high-resolution etching at the nanoscale, and is widely used in micro-nano processing, semiconductor device manufacturing and optical device processing.

3. **Process flow of electron beam etching**

The basic steps of the electron beam etching process are as follows [34, 35]:

(1) Preparation: First, the material sample to be etched needs to be prepared and mounted on the sample stage of the electron beam etching equipment. Then, set the etch parameters as needed, including etch depth, etch shape, and etch rate.
(2) Projected electron beam: An electron beam is generated from an electron beam gun and accelerated to a certain energy level by an acceleration system [30]. The electron beam is then focused to an extremely small area by a focusing system to achieve high-precision etching. Focusing systems typically consist of magnetic or electric field lenses that control the focusing and beam density of the electron beam.
(3) Etching process: After the electron beam hits the sample surface, it will interact with the atoms or molecules on the surface. This causes chemical reactions, ionization, fragmentation of atoms, or breaking of bonds, resulting in etching of the material surface. The etching process can be achieved by controlling parameters such as the energy of the electron beam, beam density, and irradiation time.

(4) Control and monitoring: In the etching process, it is necessary to ensure the accuracy and consistency of etching. Therefore, parameters such as surface topography, etch rate and etch depth can be monitored by real-time monitoring systems. These parameters can be adjusted based on real-time monitoring results to maintain the desired etch effect.
(5) Removal and post-treatment: Some by-products and residues are generated during the etching process. To keep the etched area clean and stable, by-products and residues need to be removed. This is usually achieved through a gas flow or extraction system. In addition, the etched sample can be post-processed, such as cleaning, annealing, or depositing specific materials.

The advantages of the electron beam etching process include high resolution, high accuracy, and controllability. It can be used to prepare micro-nanostructures, surface patterns and nanodevices. At the same time, because electron beam etching is not limited by light wave diffraction, it has special advantages in preparing detailed structures of high-resolution lithography. However, electron beam etching also presents some challenges, such as high cost, relatively slow speed, and limited area.

4. **Application of electron beam etching**

Electron beam etching is a key nanofabrication technology that provides an important solution for the fabrication of micro-nano devices and optical devices. Electron beam etching has a wide range of applications in micro-nano processing, semiconductor device manufacturing and optical device processing. Here are some of the main application areas:

(1) Micro-nanostructure manufacturing: Electron beam etching is one of the key technologies for the preparation of micro-nano structures. It enables high-resolution, high-precision etching for the fabrication of microscopic patterns, microfluidic channels, micromechanics, and microelectronic devices, etc. These micro-nanostructures have important application prospects in the fields of nanoelectronics, optics, and biomedicine.
(2) Semiconductor device manufacturing: Electron beam etching plays an important role in semiconductor device manufacturing. It can be used to define characteristics such as circuit shape, crystal band gap, electrodes, and channels. Electron beam etching is also widely used to manufacture devices such as integrated circuits (ICs), lithographic masks, sensors, and microelectromechanical systems (MEMS).
(3) Optical device processing: Electron beam etching can manufacture optical devices with high-precision optical surfaces. For example, it can be used to prepare gratings, diffractive optical elements, micro lenses, and photonic crystals. Electron beam etching enables the processing of precise shapes and periodic structures of optics to meet the needs of a variety of optical applications.
(4) Preparation of nanomaterials: Electron beam etching plays an important role in the preparation of nanomaterials. It can be used to prepare nanostructures, nanowires, nanodots and nanopores. These nanomaterials have a wide range of applications in nanoelectronics, catalysts, and energy storage.

(5) Research and development: The high precision and controllability of the electron beam etching process make it an important tool for the research and development of new materials, devices, and processes. It can be used to prepare samples, perform nanostructure studies, photoelectric performance adjustment, and prototype electronic devices.

In short, electron beam etching has the characteristics of high precision, high resolution, and strong controllability, and is widely used in micro-nano processing, semiconductor device manufacturing and optical device processing. With the continuous development of nanotechnology and micro-nano manufacturing technology, the application prospects of electron beam etching in various fields will be further expanded.

3.5.4 Electron Beam Additive Manufacturing

1. **Electron beam additive manufacturing principle**

Electron beam additive manufacturing is an additive manufacturing technology in which high-energy electron beams are melted, melted, and connected at the focal point created in the welded area. The basic principle is to use the energy of high-speed electron beams to melt and solidify metal materials, stacking them layer by layer to form the desired three-dimensional object. Figure 3.14 shows a schematic diagram of the electron beam additive manufacturing.

2. **Process flow of electron beam additive manufacturing**

A flow chart of the electron beam additive process is shown in Fig. 3.15. The basic methods of electron beam additive manufacturing are as follows [36]:

(1) Material preparation: select the appropriate metal powder material and pretreat it into a certain particle size and shape to facilitate the melting and melting of the electron beam.
(2) Workpiece preparation: According to the requirements of the design model, prepare the initial workpiece or supplementary materials, usually using CNC machine tools to pretreat the workpiece surface to ensure good quality and accuracy.
(3) Vacuum environment: In order to avoid the interaction and scattering of the electron beam with the gas, it is necessary to maintain a high vacuum environment in the welding area. This ensures accurate focusing of the electron beam and accurate transfer of focal energy.
(4) Electron beam melting: By adjusting the current and energy of the electron beam, a high-speed electron beam is generated in the welding area. The electron beam is focused on the focal point and shines on the layer of metal powder, causing it to heat up instantly and melt rapidly.

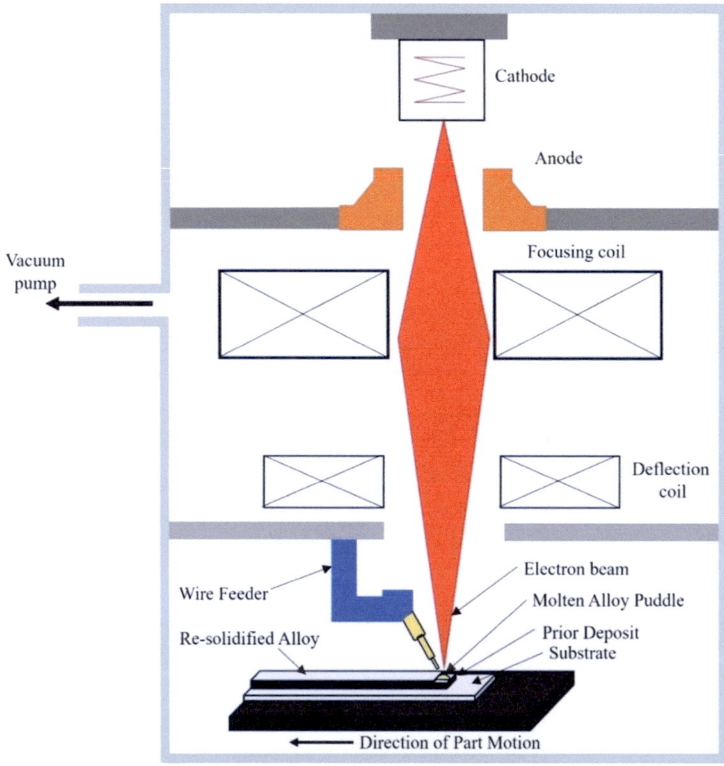

Fig. 3.14 Schematic diagram of electron beam additive material

(5) Layer by layer: The electron beam scans the movement to control the focus to move over the metal powder layer, gradually melting and stacking new metal layers. After each layer is formed, the platform descends a fixed distance so that the powder layer of the next layer is covered with the previous layer.
(6) Coating quality control: By controlling the movement and focus position of the electron beam, the uniformity, compactness and quality of the melting pool and the stacked metal layer are ensured.
(7) Cooling and solidification: After the accumulation is completed, the metal material in the welding area gradually cools and solidifies, and combines with the metal layer of the previous layer to form a whole.

3. **Application of electron beam additive manufacturing**

Through continuous layering, electron beam additive manufacturing can produce metal parts with complex shapes and high precision. This technology has the advantages of fast manufacturing speed, high manufacturing quality and high degree of design freedom, and is widely used in aerospace, automotive industry, medical equipment, and other fields [37, 38].

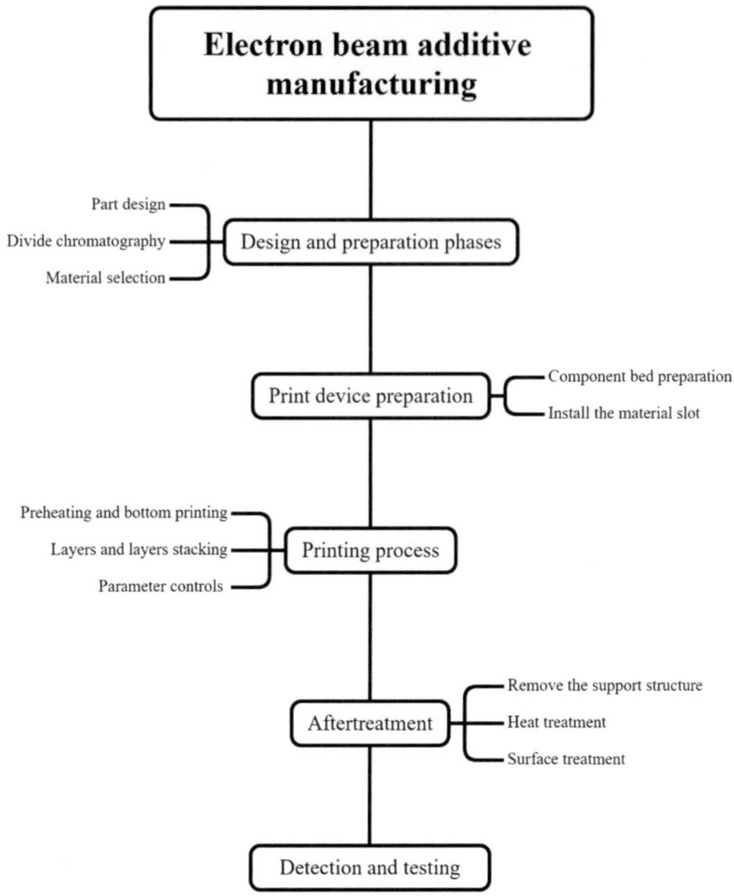

Fig. 3.15 Electron beam additive process flow chart

Additive manufacturing of electron beams as a heat source offers many advantages. First, electron beam additive manufacturing has high energy density and precise focus control capabilities, which makes it an efficient and precise manufacturing technology [39].

(1) High energy density and fast heating

Electron beam additive manufacturing uses a high-speed electron beam as a heat source with an extremely high energy density. Compared to other heat sources, such as lasers or arcs, electron beams provide a higher heat input at the same power. This means that the electron beam can heat the surface of the workpiece more quickly, allowing the metal material to melt quickly, resulting in higher manufacturing speeds.

(2) Precise focus control and high-precision manufacturing

Electron beams are retroactively manufactured by electron beams that are focused by lenses or magnetic fields to enable precise control of the position of the focus point. Compared to other heating sources, the heat-affected zone of the electron beam is more concentrated, so there is less heat loss in welding. This means that electron beam additive manufacturing can achieve greater manufacturing precision and detail control, resulting in more complex, precise parts.

(3) Wide range of material selection

Electron beam additive manufacturing has a wide range of material options. Compared to traditional heat-source additive manufacturing, electron beam additive manufacturing can handle a wide range of metal alloys, including high melting points and difficult-to-machine alloys. In addition, electron beam additive manufacturing can also handle some ceramic materials and composites, making it suitable for a wider range of applications.

(4) Low heat effect and reduce deformation

Electron beam additive manufacturing enables the control of local heating and rapid cooling by selecting the appropriate process parameters. This control reduces the size of the heat-affected area, which reduces the risk of residual stress and deformation. Compared with traditional heat-source additive manufacturing technology, electron beam additive manufacturing can better control the geometry and accuracy of parts, avoiding distortion and cracks.

(5) High temperature environment and impurity control

In electron beam additive manufacturing, the welding area can be exposed to a high vacuum, reducing contact with oxygen and other impurities. This helps control the purity of the material and can reduce the risk of oxidation and grain growth. At the same time, the high vacuum environment also helps to avoid defects such as porosity and air bubbles, improving the quality and reliability of parts.

In summary, additive manufacturing with electron beam as a heat source has the advantages of high energy density, precise focus control, high-precision manufacturing, wide range of material selection, low thermal influence, and high-temperature environment control. With the continuous development and innovation of technology, electron beam additive manufacturing is expected to play a more important role in aerospace, automotive, medical, and other fields.

3.6 Summary and Prospects

Electron beam welding technology is a high-energy-density welding method, which uses electron beams to project heat energy onto the welded joints, so that the joints are heated and melted rapidly, thereby achieving welding. Compared with ordinary welding methods, electron beam welding has the following advantages: high energy density, deep penetration welding, high-quality, high-precision welding, high efficiency and automation, excellent control performance, no contamination and so on. Due to these unique advantages, electron beam welding is widely used in the large thick parts, aerospace, electronics and instrumentation industry and production of automotive parts.

With the increasing demand for high-efficiency, high-quality and low-cost welding technology in modern manufacturing, electron beam welding technology has broad development prospects. The following is a description of the outlook for electron beam welding technology:

(1) Further improve welding quality

Electron beam welding technology can achieve high-quality welding due to its high energy density and focus focusing characteristics. However, there are still some challenges in the welding process, such as gas contamination in the welding area, residual stress, and thermal cracks, etc. Future research will focus on developing new shielding gases and substrate treatments to reduce residual stress and improve joint quality, thereby further improving welding quality.

(2) Expand the range of materials

Electron beam welding technology is currently widely used in the welding of metal materials such as iron, steel, and titanium alloys. In the future, as the demand for welding of non-metallic materials increases, electron beam welding technology will begin to be applied to the welding of non-metallic materials such as plastics, composite materials, and ceramics. In order to realize the welding of these non-metallic materials, it is necessary to develop suitable beam current parameters, surface treatment methods and heating control techniques.

(3) Improve production efficiency

Electron beam welding technology has high welding speed and heat input control ability due to its high energy density and local heating characteristics. Future research will focus on increased welding speeds and rapid temperature control methods to achieve higher productivity. At the same time, it is also necessary to further reduce equipment costs and increase the degree of automation to meet the needs of mass production.

(4) Develop multifunctional electron beam welding

Electron beam welding technology has the advantages of high energy focus, low heat input and small heat affected zone, which enables it to realize the integration of other processing functions. Future research will focus on combining electron beam welding with other processing methods (e.g., nanofabrication, powder spraying) and other electron beam processing methods (e.g., cutting, surface modification, ectching, additive manufacturing).

Problems

1. What is electron beam welding?

2. How does electron beam welding work?
3. What are the limitations of electron beam welding?
4. How is electron beam welding applied in the fabrication of large and thick components?
5. How is electron beam welding used in the aerospace industry?
6. What specific advantages does electron beam welding offer in aerospace applications?
7. How does electron beam welding contribute to the electronics and instrument industry?
8. What advantages does electron beam welding provide in electronics and instrument manufacturing?
9. What are the benefits of using electron beam welding in automotive applications?
10. How is electron beam welding utilized in additive manufacturing?
11. How is electron beam cutting different from other cutting methods?
12. What are the advantages and disadvantages of electron beam cutting?
13. What types of surface modifications can be achieved using electron beam technology?
14. How does electron beam surface modification enhance material properties?
15. What are the applications of electron beam etching in various industries?
16. How does electron beam etching compare to other etching methods?
17. What are the specific applications of electron beam technology in semiconductor manufacturing?

References

1. Sayegh G, Nazarenko OK, Dietrich W et al (2007) An international history of electron beam welding. Pro-beam AG Germany, pp 1–20
2. Meleka AH (1971) Electron beam welding: principles and practice. McGraw-Hill, New York
3. Schulze TG (2012) AS23-04—the phenotype—the still unknown quantity in psychiatric genetics? Eur Psychiatry 27:1
4. Weglowski MS, Błacha S, Phillips A (2012) Electron beam welding—techniques and trends. In: Chaturvedi M (ed) Welding and joining of aerospace materials, 2nd edn. Woodhead Publishing, London, pp 157–198
5. Reisgen U, Gach S (2021) Chapter 2—electron beam welding. In: Da Silva L, El-Zein M, Martins P (eds) Advanced joining processes. Elsevier, New York, pp 35–65
6. Węglowski MS, Błacha S, Phillips A (2016) Electron beam welding–techniques and trends–review. Vacuum 130:72–92
7. Arata YT et al (1973) Study on characteristics of weld defect and its prevention in electron beam welding (report I): characteristics of weld porosities. Trans JWRI 2(1):103–112
8. Wang X, Wang H, Ma J et al (2023) Numerical investigation of the impact of thermoelectric effect on electron beam welding of dissimilar materials. J Market Res 27:5315–5320
9. Antony K, Arivazhagan N (2015) Studies on energy penetration and Marangoni effect during laser melting process. J Eng Technol 10(4):509–525
10. Węglowski MS, Dworak J, Błacha S (2014) Electron beam welding–characteristics. Biuletyn Instytutu Spawalnictwa w Gliwicach 58(3):5–14

11. Xian DG (2021) Research on assembly and welding methods of large structural devices. Public Stand 19:19–21 (in Chinese)
12. Xia XW, Wu JF, Liu ZH et al (2022) Numerical simulation of 50 mm 316L steel joint of EBW and its experimental validation. Metals 12(5):725
13. Chen GQ, Shu X, Liu JP (2018) Development status of applications of vacuum electron beam welding technology. J Netshape Form Eng 10(1):31–39 (in Chinese)
14. Lei SL (2014) Discussion on light alloy welding process in aerospace. Technol Enterp (22):143 (in Chinese)
15. Wencel Z, Wiewiórowska S, Wieczorek P et al (2023) Electron beam welding process for Ti_6Al-4V titanium alloy. Materials 16(14):5174
16. Quan ZN, Bai HB (2013) Study on welding technology of liquid cooling board for airborne electronic equipment. Hot Work Technol 42(15):224–225 (in Chinese)
17. Feng X (2011) Structure design of welded cabinet for shipborne electronic equipment. Electro Mech Eng 27(4):16–20 (in Chinese)
18. Jing B, Hu JX, Huang YF et al (2016) Evaluation and prospect of thermal fatigue on lead free solder joints in electronic equipment. J Air Force Eng Univ 17(6):35–40 (in Chinese)
19. SJ/T 10414-2015. Solder for semiconductor device. Ministry of Industry and Information Technology of the People's Republic of China, 2015 (in Chinese)
20. Zhang XQ, Li JS, Tao MF et al (2022) Electron beam welding and post heat treatment of a new near-beta high-strength Ti-4Al-5Mo-5V-5Cr-1Nb alloy. Metals 12(8):1396
21. Wan T, Xi DX, Wang WJ (2022) Process flow and integrated control technology of automobile welding production line. Manuf Autom 44(1):68–70 (in Chinese)
22. Chen Y, Zhai FX (2021) Application of welding automation technology in automobile industry. Intern Combust Engine Parts 2:185–186 (in Chinese)
23. Noga P, Skrzekut T, Wędrychowicz M et al (2023) Research of friction stir welding (FSW) and electron beam welding (EBW) process for 6082–T6 aluminum alloy. Materials 16(14):4937
24. Li HX, Lin F (2022) Research on electron beam gasification cutting process for electron beam addition and subtraction technology. Electromachining Mould 5:53–63 (in Chinese)
25. Hou J (2009) Study on microstructure of TC4 titanium alloy high-energy beam welding weld. Huazhong University of Science and Technology, Wuhan (in Chinese)
26. Zhang WL, Sun T, Li JY (2003) The smelting of electron beam and the smelting equipment. Metall Equip 4:32–34, 31 (in Chinese)
27. Chen YF, Xian Y, Jin TY (2009) Electron beam engineering technique and application. Mod Manuf Eng 8:153–156 (in Chinese)
28. Gong Y (2021) Research on surface strengthening treatment of 40cr steel by scanning electron beam. Guilin University of Electronic Science and Technology, Guilin (in Chinese)
29. Zhao H, Wang JC, Du CY et al (2018) Research on electron beam surface modification. J Shenyang Ligong Univ 37(3):47–52 (in Chinese)
30. Wei DQ, Li XK, Ren XL (2016) Research progress on surface modification of electron beam. Hot Work Technol 45(18):19–23 (in Chinese)
31. Hao Y (2012) Influence of intensive pulsed electron beam treatment on surface microstructure and properties of hypereutectic Al-Si alloy. Northeastern University, Shenyang (in Chinese)
32. He F, Zhang Y, Lü HB et al (2017) Analysis of the causes of high-voltage discharge on EB-PVD. Vacuum 54(5):10–14 (in Chinese)
33. Huang K (2018) Research on physical design of extremely cold electron source. University of Science and Technology of China, Hefei (in Chinese)
34. Yang Y (2007) Application of surface nano needle array technology in Pd PPF lead-free lead frame. Shanghai Jiaotong University, Shanghai (in Chinese)
35. Wang Y (2021) Optimum design of simulation and optimal design of wide scanning electron beam dynamic focusing. Harbin Institute of Technology, Harbin (in Chinese)
36. Zhang GD, Zhang P, Gao JS et al (2023) Microstructure and mechanical properties of TC11 titanium alloy fabricated by wire-feed electron beam additive manufacturing. J Mech Eng 59(4):105–112 (in Chinese)

37. Feng N (2021) Optimal design of composite material structure for residual vibration suppression. Dalian University of Technology, Dalian (in Chinese)
38. Wu F, Lin BC, Quan YZ et al (2022) Review on equipment and application of electron-beam based additive manufacturing. Vacuum 59(1):79–85 (in Chinese)
39. Li S, Gao Q, Zhao J (2021) Research progress and prospect of electron beam freeform fabrication. Mater China 40(02):130–138 (in Chinese)

Chapter 4
Friction Stir Welding

4.1 Principle, Characteristics and Classification of Friction Stir Welding

4.1.1 Principle of Friction Stir Welding

The friction stir welding (FSW) technique was first invented by Wayne Thomas and E. Nicolas at The Welding Institute (TWI) of the United Kingdom (UK) in 1991. FSW uses a rotating tool that joins two separate specimens. The specimen body is subjected to rotational or linear oscillation, and material fusion occurs at the interface through frictional heat. FSW is a very popular and efficient metal joining process that produces high-quality welds for difficult-to-weld low-density ductile metals and their alloys (e.g. aluminum, magnesium, titanium alloys), as well as polymer composites. Furthermore, FWS is a very flexible welding technique capable of joining two different materials such as copper, zinc, magnesium, steel, titanium, nickel and their alloys. Consequently, FSW has become the most popular choice for welding light ships, deck panels, hulls, bodies, frames, fuel tanks, aircraft wing structures, electronics, and circuits. However, FSW also has disadvantages, such as complex fixture setup for clamping workpieces, large downward force for tool insertion, and lower flexibility than manual and arc welding processes [1, 2].

The schematic of FSW is illustrated in Fig. 4.1a. It starts with the rotation of a cylindrical friction stir tool which is driven by an electric motor and consists of two components: a probe and a shoulder. Shoulder rotation creates friction at the interface of the work-piece and the tool material, generating heat energy sufficient enough to soften the material at the junction. The probe is completely shoved into specimen thickness steadily maintaining the contact of the shoulder with the metal specimen. Preventing molten material from being thrown out of the weld region is also one of the functions of the tool shoulder [3, 4]. Once sufficient heat is produced, the tool is traversed in the direction of the joint line, and the softened material then

R. Li and T. Li, *Advanced Welding Methods and Equipment*,
https://doi.org/10.1007/978-981-97-4109-0_4

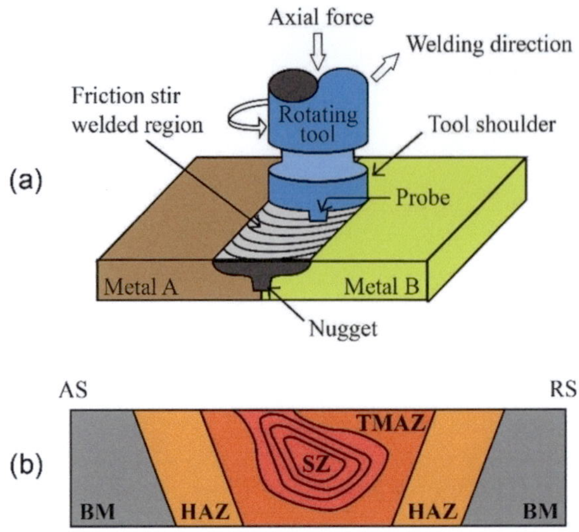

Fig. 4.1 Schematic of FSW (**a**) and its weld zones (**b**)

simultaneously gets displaced and stirred with the spinning effect of the probe [5]. Moving across the weld line, the rotating tool extrudes the fused material in a specific flow pattern and forges the materials together, leaving behind the plasticized material coalescing to seal the joint. Although the specimen gets heated up, the temperature never reaches the melting point of the specimen throughout the process. This eliminates the problems regarding re-solidification such as weld porosity, embrittlement, solute redistribution, and solidification cracking. A very fine equiaxed re-crystallized grain structure is produced, bearing similar properties to those of the parent materials. The versatility of the FSW process shows its applications in all weld geometries, namely lap, butt, T, and fillet joints [6–8].

The weld zones of FSW are shown in Fig. 4.1b, where AS stands for the advancing side and RS stands for the retreating side. The friction heating and plastic deformation generated by the rotating and stirring effect of the FSW tool form different zones termed as the base material (BM), heat affected zone (HAZ), thermo-mechanically affected zone (TMAZ), and stir zone (SZ) which is frequently termed as the nugget zone (NZ). Each weld zone exhibits distinct microstructural characteristics that determine the quality and properties of the welded joint. TMAZ is located between HAZ and SZ; the grains of TMAZ have retained the microstructure but in an elongated and deformed state. The most severely deformed and being dynamically re-crystallized zone is the stir zone, which has the lowest hardness value among the rest of the weld zones despite its equiaxed grain structure. HAZ is known as the weakest zone as the materials in this zone are thermally affected and, therefore, for enhancement in the mechanical properties of this zone, FSW is conducted underwater to control temperature levels. The materials in TMAZ are plastically deformed and are thermally affected as well, whereas in HAZ the materials are not plastically deformed.

Recrystallization and elongation of grain structure which are accountable for the improvement in the tensile strength of the joint were observed to be initiated in TMAZ [9].

4.1.2 Characteristics of Friction Stir Welding

Friction stir welding has the following features shown in Fig. 4.2.

No-melt welding: Friction stir welding is a solid-state welding technique in which the welding process does not require heating to the melting point of the material or the addition of filler material. The weld area is only softened by localized heating and mixed together by stirring to achieve a weld.

Mixed material: Friction stir welding applies stirring force and frictional heat to the weld area by rotating the tool head, which softens and mixes the material to be welded together to form a continuous, uniform weld. This mixing action helps eliminate defects in the original material and improves the strength of the weld.

High welding speed: Friction stir welding offers high welding speeds compared to conventional welding methods. Since there is no need to wait for melting and solidification during the welding process and there is no restriction on filler material, fast and continuous welding is possible.

Good mechanical properties: Friction stir welding joints have high strength and toughness. As the welding process maintains the fibre structure and grain morphology of the original material, there will be no grain growth, solid solution precipitation or other problems, thus improving the mechanical properties of the welded joints.

Wide range of application: Friction stir welding is applicable to various metals and alloys, including aluminum alloys, titanium alloys, magnesium alloys and so on. Compared with traditional arc welding, friction stir welding is more tolerant to changes in material composition and hardness.

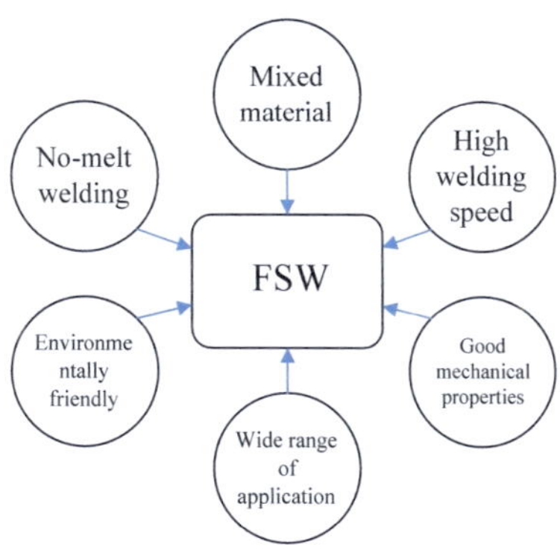

Fig. 4.2 Features of friction stir welding

Environmentally friendly: Friction stir welding does not require the use of a welding arc or other high-temperature heat source, and therefore does not produce welding fumes, exhaust fumes or radiation. It is a relatively environmentally friendly welding method.

However, there are some limitations to Friction stir welding.

High equipment costs: Friction stir welding requires specialized equipment and tools, including rotary heating tool heads and stirring devices. These are relatively expensive to purchase and maintain.

Limitations on weld thickness: Friction stir welding is suitable for welding medium and thin plates. For thicker materials, multiple passes or other auxiliary welding methods may be required.

High demands on workpiece shape: Friction stir welding requires the workpiece to be held in a specific position, which places certain restrictions on the shape and size of the workpiece.

These features make friction stir welding a welding technology with a wide range of prospects for application in aerospace, automotive manufacturing, shipbuilding, rail transport and other fields.

4.1.3 Classification of Friction Stir Welding

According to different applications and welding methods, friction stir welding can be categorized into the following classifications shown in Fig. 4.3.

Linear Friction Stir Welding (LFSW): This is the most common type of stir friction welding. The material is softened and mixed together by rotating a heated tool head that moves across the weld.

Conical Friction Stir Welding (CFSW): In contrast to linear friction stir welding, conical friction stir welding has a conical shaped tool head. This method is suitable for thinner materials and applications requiring higher weld strength.

Friction Stir Spot Welding (FSSW): This method uses a friction motion similar to stirring, but with a smaller weld area for applications requiring localized welding.

Rotational Friction Stir Welding (RFSW): This welding method uses a rotating tool head and applies an axial force to force the material to rub against the weld.

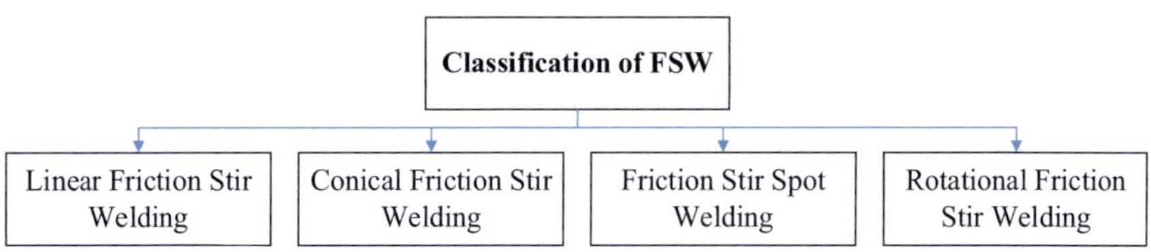

Fig. 4.3 Classification of FSW

These classifications only represent some of the variants of friction stir welding. Depending on the specific application requirements and material properties, other types of friction stir welding are possible.

4.2 Parameters of Friction Stir Welding

Tool rotational speed, welding speed, vertical pressure on the tool, and tilt angle of the tool are the main independent variables that are used to control the FSW process. These process parameters (shown in Fig. 4.4) define the structural and mechanical characteristics of the friction stir welded joints and are related to the weld defect phenomena as well. Therefore, to achieve superior quality welded joints, the influence of the process parameters during the FSW process needs to be understood thoroughly.

4.2.1 Tool Rotational Speed

The rotational speed of the tool during FSW influences heat generation at the surface of the specimen, significantly affecting plastic deformation, material flow, and the grain structure of the material at the weld zone. Higher tool rotation rates generate higher temperature because of higher friction heating, and result in more intense stirring and mixing of material. However, it should be noted that the frictional coupling of the tool surface with the workpiece is going to govern the heating. So, a monotonic increase in heating with increasing tool rotation rate is not expected as the coefficient of friction at the interface will change with increasing tool rotation rate.

Tool rotational speed is the most influencing factor for the attainment of high tensile strength along with the tensile elongation. According to the FSW study of AZ31 magnesium alloy [10], for a 2.5 mm thick workpiece, the joint strength without any treatment after welding can reach 93% of that of the base metal. The best welding parameters are a welding speed of 95 mm/min and a rotation speed of 1180 r/min. For

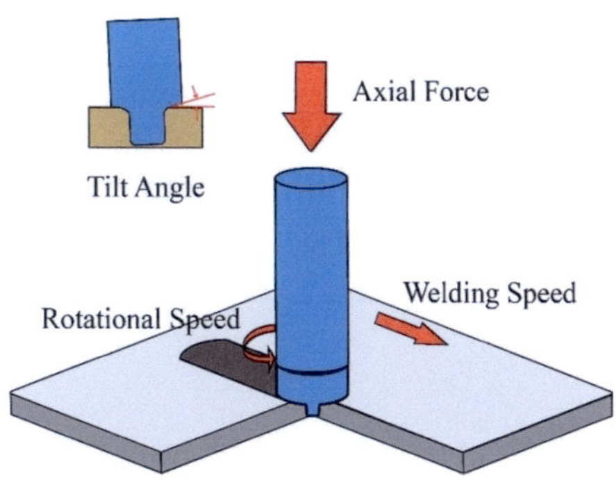

Fig. 4.4 Parameters of FSW

a 5 mm thick magnesium alloy, the joint strength nearly equals that of the base metal. The best welding parameters are a welding speed of 150 mm/min and a rotation speed of 800r/min. The fracture locations are in the TMAZ. Small porosities are easy to occur when using unreasonable parameters and these small pores grow, gather, and finally fracture at the disturbing of external force during tensile tests.

4.2.2 Welding Speed

Welding speed, sometimes also termed as the traverse/moving speed of the FSW tool along the direction of the joint line. Generally, the welding temperature increases with the decrease of welding speed, and the tensile strength of the friction stir welded joints decreases with the increase of the welding speed. For FSW of Al-Mg alloys, lower welding speed leads to high temperature and the formation of liquid films at aluminum–magnesium alloys interface can be observed. The formation of liquid films leads to tool slippage and reduced friction heating which further results in cracking defect and development of brittle IMCs such as Al_3Mg_2 and $Al_{12}Mg_{17}$ after solidification, significantly weakening the joint strength [11]. Reducing the welding speed while increasing the speed of tool rotation with high axial pressure can avoid weld zone defects such as void, cracking, or distortion, with further improvement in the strength of friction stir weld joint [12]. Increasing welding speeds increases the transverse tensile properties of 7050-T7451 aluminum alloy plates with reduced grain size in the SZ and TMAZ. Whereas, decreasing welding speeds results in the widening of the nugget zone, reducing the hardness of friction stir welded joint [13].

For FSW of aluminum matrix composite, the pinhole defect observed at the joint might be due to the combined set of low values of rotational speed and traverse speed with insufficient heat generation and inadequate material transfer. However, a tunnel hole defect might be formed due to excess heat and insufficient heat consolidation caused by combined high values of rotational and traverse speed. Figure 4.5a shows the polishing sections of FSWed samples of AA5083/AA5754 at different rotational and traverse speeds[14]. Arrows indicate the tunneling defects formed in the welded samples. It shows that high heat input results in tunneling defects. Figure 4.5b shows the FSWed samples of AA5083/AA7020 at different rotational and traverse speeds. It is observed that the nugget zone is wide and conical at the top surface because of the large shoulder diameter and narrow cylindrical at the bottom surface due to the small pin. On the AA5083 side, increasing the traverse speed contributes to the formation of tunneling defects. Due to the high resistance of AA7020, the interface is free of defects irrespective of the position of the alloy on the advancing or retreating side.

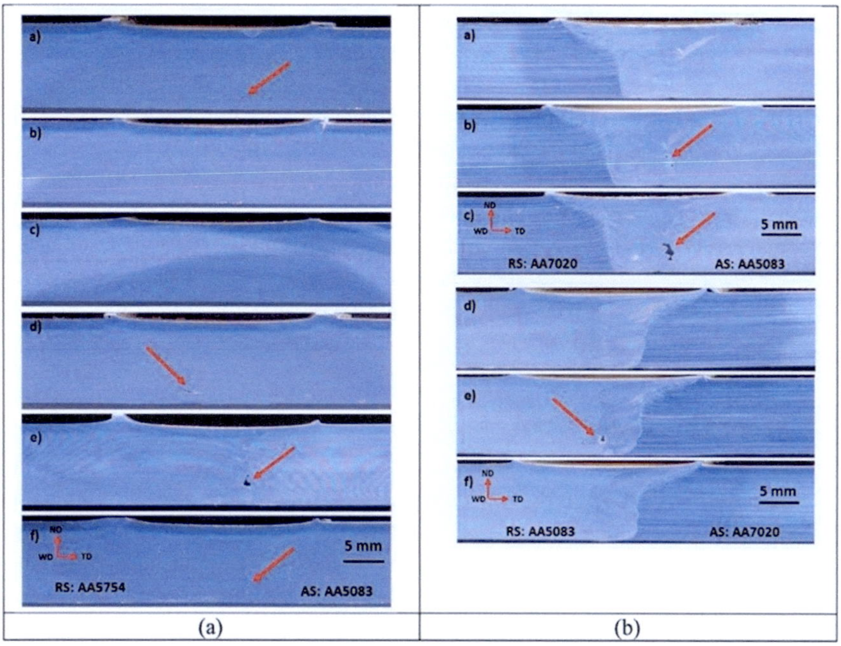

Fig. 4.5 Macro images of FSWed samples of **a** AA5083/AA5754 and **b** AA5083/AA7020 at different rotational and traverse speeds [14]

4.2.3 Tilt Angle and Axial Force

In addition to the tool rotation rate and traverse speed, another important process parameter is the angle of spindle or tool tilt with respect to the workpiece surface. A suitable tilt of the spindle towards the trailing direction ensures that the shoulder of the tool holds the stirred material by the threaded pin and moves material efficiently from the front to the back of the pin. The general tile angle is 2°–5°.

Generally, the axial force or welding pressure changes with the inserting depth. Therefore, the insertion depth of the pin into the workpieces (also called target depth) is important for producing sound welds with smooth tool shoulders. The insertion depth of the pin is associated with the pin height. When the insertion depth is too shallow, the shoulder of the tool will not contact the original workpiece surface. Thus, the rotating shoulder cannot move the stirred material efficiently from the front to the back of the pin, resulting in the generation of welds with the inner channel or surface groove. When the insertion depth is too deep, the shoulder of the tool will plunge into the workpiece creating excessive flash. In this case, a significantly concave weld will be produced, leading to local thinning of the welded plates. The axial force has a significant influence on the grain size in the weld zone, the formation of weld defects, and hardness of the weld in SZ, which subsequently affects the weld properties. The high axial force causes temperatures to rise in the nugget zone with an increased flow of plasticized material, while lower axial force results in the generation of insufficient friction heat and the forging pressure [15]. For FSW of magnesium alloys (AZ61A) in Razal Rose's study, application of 5 kN axial force revealed enhancement in tensile strength and the hardness numbers a result of finer grain structures in the SZ [16]. The elongated grain structure of the base material 7020-T6 aluminum alloy is improved to fine re-crystallized equiaxed grains by increasing axial load. Increased axial load also results in the defect-free welded joints with high strength efficiency compared to the base material [17]. However, excessive axial pressure can be a significant indicator of tool erosion, which in extreme cases can lead to tool breakage, and excessive pressure can also lead to overheating and thinning of the joint.

4.3 Tool Designs of FSW

FSW tool consists of a cylindrical rotating shoulder with a threaded tool pin called a probe. Tool functioning includes heat generation, blending and stirring of materials, cracking of the joint line, dispersion of oxide layers, developing forging pressure, prevention of metal surface from weld flash, and eliminating defects such as wormholes, sheet-thinning, or hooking defects. Weld quality and tool wear are two important aspects taken into consideration while selecting a tool material. Properties of the tool material namely tool strength, fracture toughness, hardness number, thermal conductivity, and coefficient of thermal expansion are the major regulating parameters for the tool performance [18–20]. The strength of the work-piece material determines the stresses induced in the tool. Chemical exposure or oxidation at the tool surface during or after plunging also causes tool erosion which is controlled by providing a layer of chromium or titanium coating. The tip is typically made of polycrystalline cubic boron nitride (PCBN), or polycrystalline diamond (PCD), due to their characteristics of extremely high-temperature tolerances, hardness, and abrasiveness [2].

4.3.1 Tool Geometry

Tool geometry has been significantly taken into consideration while designing a tool for the FSW process as it affects the input energy, deformation patterns, plunging force, microstructure, and subsequently the mechanical properties of joints. Moreover, the formation of weld defects is significantly influenced by the tool geometry. Shoulder geometry and pin geometry are the main contents of tool geometry. Tool geometry has a significant influence on the FSW process and the main tasks of the tool are as follows:

(1) Heat the workpiece
(2) Induce material flow

(3) Restrict the heated metal below the tool shoulder.

Tool shoulders are devised to frictionally heat the surface area of the workpiece. The flat shoulder is the simplest, but it produces excessive flash due to the inability of the flat shoulder end surface to trap the flowing material. The concave shoulder has been proven to be a better solution as it restricts material from both sides of the shoulder. Tool shoulder diameter generates most of the heat and contacts with plasticized materials. It controls the material flow field which further determines the quality of the weld. The designing criterion for the shoulder diameter of the tool is based on torque usage computed by the heat transfer rate and flow rate of the material [21]. Larger diameter significantly increases the amount of heat generation at the surface reducing the stresses [22]. Table 4.1 summarizes several different types of FSW tool shoulder geometry with their influence on weld joint properties and drawbacks [23, 24].

Several different tool pin geometries are illustrated in Fig. 4.6. Among the various tool geometries, threaded cylindrical, triangular probe and square configured geometries are widely used in FSW fabrication [7]. The shape of the tool influences the flow of material and the weld properties. Threaded tool profiles give better results in comparison to none threaded tool profiles. However, square pin profiles yield better mechanical and metallurgical properties [14]. Triangular pins increase the material flow more than cylindrical profiles, and a triangular prism-shaped tool pin is a good choice for harder alloys. Threaded profiles give full consolidated weld, while non-threaded profiles produce voids. Tool profiles affect the weld quality, tool rate and homogeneous grain structure. Pin and shoulder diameter are two important factors

Table 4.1 Different types of FSW tool shoulder [23]

Types of tool shoulders	Flat	Concave	Convex	Conical	Scrolled
Geometry	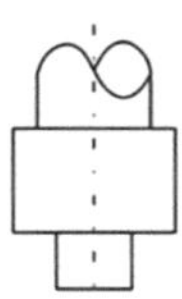	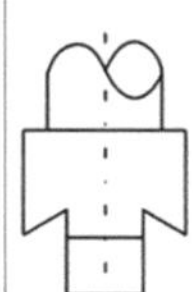	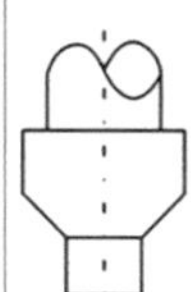	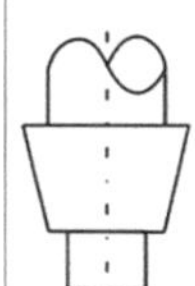	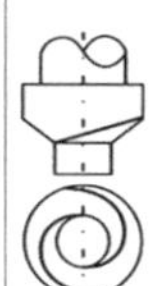
Advantages	Requiring low spindle torque	Good material flow, improved nugget zone integrity	Ability to weld complex surfaces and suitable for gauge thickness less than 1 mm	Excellent weld surface appearance	Providing superior material flow, grain refinement and high tensile-shear strength
Drawbacks	Poor material flow, excessive material flash	Sometimes requiring tilting of the tool	Poor weld surface finish and weld integrity	Reduced thickness of weld surface	Rough weld surface appearance

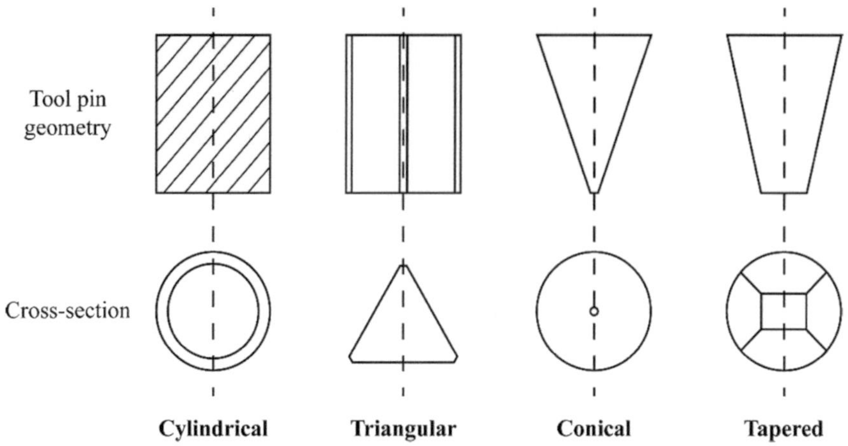

Fig. 4.6 Tool pin geometry with their cross-section

to consider when designing a tool. The heating area increases accordingly with the diameter of the shoulder. A threaded tapered pin used for butt welded joint of AA6061 aluminum alloy result in high heat generation with improved stirring effect, a defect-free weld joint of hardness value 95% to that of the base material is obtained [25, 26]. Cylindrical pin geometry employed for welding aluminum alloy (AA2198-T4) in lap joint provides large nugget area constraining the upward flow of the plasticized material caused by tool translation, while conical pins serve the smaller loadbearing area with uniform temperature distribution which further helps in reducing distortion [25, 27]. The triangular pin tool effectively increases the material flow compared to the cylindrical pins and reduces the hook height on the retreating side consequently improving the strength of the weld joint [28, 29].

4.3.2 Tool Materials

High-Speed Steel (HSS) is a widely used tool material for FSW, for welding soft materials like aluminum, copper, and their wide range of combinatorial alloys. In the case the welding of steels, HSS is ineffective as it cannot withstand high temperatures and abrasive wear. Pure tungsten and polycrystalline cubic boron nitride tools are being explored. Pure tungsten has poor toughness at ambient temperature, and rapid wear may takes place during welding. Tungsten carbide (WC) has been explored as an alternative due to its high hardness of approximately 1650 HV and ability to withstand high loads during welding trials. For FSW of steels, Polycrystalline Cubic Boron Nitride (PCBN) is the best option due to its hardness and ability to withstand high temperatures [30]. Table 4.2 summarizes frequently used tool materials employed during FSW with their weld properties.

Table 4.2 Different tool materials used for FSW

References	Work-piece material	Tool material	Tool geometry	Weld properties
Matsushita et al. [31]	Carbon nano-tube reinforced aluminum matrix composites	Tool steel HI3	Flat	Refined grain size, increased tensile and yield strength
Matsushita et al. [31]	Ferritic-martensitic stainless steel	Polycrystalline cubic boron nitride (PCBN)	Concave	High hardness and strength of SZ achieved with nearly 100% joint efficiency
Li et al. [32]	6013-T4 aluminum alloy	H15 steel	Concave	Tensile strength of the weld is 70% to that of the BM
Kumar et al. [33]	AA6063 and AA5052 aluminum alloys	Ledeburite alloy steel	Modified tool with circular walled structure	Average tensile strength of 159 MPa achieved
Fu et al. [34]	AA6061-T4	H13 steel	Tapered probe	Fine and equiaxed grain with joint efficiency up to 80%
Mofid et al.[35]	6061-T6 aluminum alloy and AZ31B-O magnesium alloy	H13 tool steel	Concave shoulder with conical threaded pin	Tensile strength up to 70% of the magnesium BM

4.4 Friction Stir Welding of Various Materials

4.4.1 Friction Stir Welding of Aluminum Alloys

To date, the main focus of friction welding has been on welding aluminum alloys. Considerable work has focused on the use of friction welding to join different aluminum alloys. The steady push into light vehicles is largely responsible for the research into linking aluminum. Because friction welding is a solid state process, it can be used to join all common aluminum alloys, including the 2xxx, 7xxx, and 8xxx series, which are often challenging or impractical to weld through the fusion process. The weldability of different aluminum alloys is shown in Fig. 4.7. In work hardened alloys (e.g. 5xxx), the heat of the friction welding process will allow heat recovery and recrystallization of the dislocation substructure, although this is partially offset in the intensely deformed region that produces the new dislocation structure. In age-hardened alloys, the weld is often heated to a dissolution temperature well above that

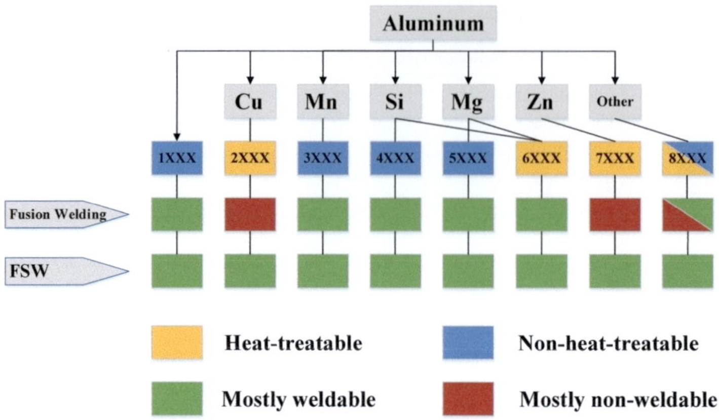

Fig. 4.7 Weldability of different aluminum alloys

of the initial precipitate, resulting in dissolution, reprecipitation, and overaging. As a result, friction stir welded aluminum alloys will contain microstructures covering the entire range of normal temper.

The advantages of FSW for welding aluminum can be summarized as follows.

(1) As a solid state process it can be applied to all the major aluminum alloys and avoids problems such as hot cracking, porosity and element loss common to aluminum fusion welding processes.
(2) As a mechanized process, FSW does not rely on specialized welding skills; indeed manual intervention is seldom required.
(3) No shielding gas or filler wire is required for aluminum alloys.
(4) The process is remarkably tolerant to poor quality edge preparation: gaps of up to 20% of plate thickness can be tolerated, although this leads inevitably to a reduction in local section thickness since no filler is added.
(5) The absence of fusion removes much of the thermal contraction associated with solidification and cooling, leading to significant reductions in distortion; however, it is not a zero distortion technique.
(6) It is very flexible, being applied to joining in one, two and three dimensions, being applicable to butt, lap and spot weld geometries; welding can be conducted in any position.
(7) Excellent mechanical properties, competing strongly with welds made by other processes.
(8) Workplace friendly: there are no ultraviolet or electromagnetic radiation hazards as the absence of an arc removes these hazards from the process; the process is no noisier than a milling machine of similar power, and generates virtually zero spatter, fume and other pollutants.
(9) The energy required at the weld for FSW lies between laser welding (which requires less energy) and GMAW (gas metal arc welding) (which typically needs more).

(10) High welding speeds and joint completion rates: in single pass welds in thinner materials (down to 0–5 mm thickness), FSW competes on reasonable terms with fusion processes in terms of welding speed; in thicker materials, FSW can be accomplished in a single pass, whereas other processes need multiple passes. This leads to higher joint completion rates for FSW, even though the welding speeds may be lower. Thick plates can also be joined by FSW on either side.

(11) Various mechanical and thermal tensioning strategies can be applied during welding to engineer the state of residual stress in the weld.

There are of course disadvantages to FSW; indeed, some of the advantages listed above can be viewed in a less positive light in certain circumstances. For example, the absence of a filler wire means that the process cannot easily be used for making fillet welds. Similarly, the fully mechanized nature of the process prevents its use for applications where access or complex weld shape is best suited to a manual process. The presence of a hole at the end of the weld from which the probe was withdrawn is often quoted as a disadvantage. In practice, this has seldom been a significant problem, as there are many possible solutions, which have been considered elsewhere. The workpiece also needs to be restrained in well-designed support tooling both to react to the forces applied, and to prevent the probe from pushing the workpiece materials apart. Although the process may reduce the strength of aluminum alloys, this can be compensated for if necessary by appropriate design of the joint, for example by locally increasing the thickness, but in most cases no changes are made. The process economics of FSW are generally considered favorable, but specific published data are lacking. However, it is known that the process drastically reduces weld preparation costs, skilled welder requirements and repair rates.

Friction stir welds in aluminum alloys contain a wide variety of microstructures, which is not surprising when considering the extreme strains, strain rates, and thermal cycle ranges that different regions of the weld are subjected to. In HAZ far from the center of the weld, there is no significant change in grain structure, and HAZ is detected only by changes in hardness and changes in etching response [36]. In precipitate-hardened alloys, it is generally accepted that some coarsening is taking place in the sediments and that they may dissolve at higher temperatures. In work-hardened alloys, the dislocation network may recover, which may result in the formation of some low-angle cell boundaries. In addition, when near the center of the weld, clear evidence of plastic deformation can be seen in the grain structure. Outside the TMAZ, the original grains are still recognizable in the deformed structure, and the formation of sub-grain structures and the rotation of the parent grains are significantly correlated. Closer to the fuse line, the strain, temperature and duration of high temperatures increase, resulting in the formation of recrystallized nuggets with a fine equiaxed structure.

Table 4.3 [37] shows the mechanical properties of friction welded joints of 5xxx, 6xxx and 7xxx series aluminum alloys. It is found that the tensile strength of the 5083-O aluminum alloy joint is similar to that of the parent metal. After solution treatment and artificial aging treatment, the tensile strength of 6082 aluminum alloy

Table 4.3 Properties of friction stir welding joints of typical aluminum alloys [37]

Alloys	Category	Yield strength (MPa)	Tensile strength (MPa)	Elongation (%)	Strength coefficient (%)
5083-O	Welded	142	298	23	100
	Parent metal	148	298	23.5	–
6082-T6	Welded	160	254	4.85	83
	Parent metal	286	301	10.4	–
	Welded + Aging	274	300	6.4	100
7108-T79	welded	210	320	12	86
	Parent metal	295	370	14	–
	Welded 1 Natural aging	245	350	11	95
7075-T7351	Welded	208	384	5.5	70
	Parent metal	476	548	13	–

friction welded joint also reached the strength of the parent metal, and the elongation decreased after heat treatment. The tensile strength of 7108 aluminum joint is 95% of that of base material after natural aging at room temperature. The fatigue test used 5083-O aluminum alloy welds with a thickness of 6 mm. When the stress ratio is set to 0.1 (R = 0.1), the fatigue property of 5083-O aluminum alloy weld is similar to that of the parent metal. A large number of experimental results show that the fatigue performance of friction welded butt joint exceeds the recommended design value of fusion welded joint. The fatigue test results show that the fatigue performance of friction weld is the same as that of fusion welded joint. However, in most cases, the fatigue properties of friction welds are higher than those of fusion welds.

4.4.2 Friction Stir Welding of Magnesium Alloys

FSW of magnesium alloy involves AZ, AM, and AC series, such as AM50, AM60, AZ91, AZ61, and AZ31. The strength coefficient of magnesium alloy FSW joints could reach 90% or even 100%. The well-formed joints can be obtained at the rotation speed of 600–1000 rpm and the welding speed of 600–1000 rpm. The FSW of hot-rolled AZ31 magnesium alloy was further investigated by the Shibaura Institute of Technology in Japan. The results indicated the strength of the joint could be comparable to that of the parent alloy. Moreover, the mechanical properties of the joints were closely related to the state of the parent metal. The properties of typical magnesium alloy FSW joints are listed in Table 4.4 [37]. Input parameters result in changes in the microstructural evolution and mechanical properties of friction-stir welding of Mg alloys. The use of FSW of the Bobbin tool (BTFSW) on alloy ZK60 Mg resulted in significant grain refinement and disintegration of the precipitation zone (SZ) Mg_4Zn_7. Significant thermal softening at the ZS resulted in a lower yield strength of the BTFSW joints than that of the BM. This is mainly attributed to grain refinement influenced by the welding speed of 300–400 mm/min. Similarly, a carefully selected rotational speed of 300–800 rpm and a constant tool speed of 500 mm/min using a ZK60 Mg alloy-controlled FSW type. The UTS of the joint was 91–95% of the BM. The input parameters allowed the microstructure to be refined with Zr-rich precipitates and bimodal recrystallized grain sizes ranging from 0.6 to 1.5 mm in the SZ. Dissolution of Mg-Zn precipitates at SZ due to welding speed was accompanied by significant thermal softening. The effect of heat input and rotational speed of FSW seals of alloy AZ61 Mg has been studied [38]. It has been determined that heat input is proportional to welding speed and indirectly proportional to rotation speed. A welding and rotation speed of 2 is not desirable for AZ61 alloy joints due to the formation of tunnel, cavity and flash defects in the weld region. The grain refinement of the SZ was attributed to the reduction in the speed of rotation of the tool and vice versa. This is also supported by other Mg alloys. Improper penetration, bound kisses, tunnel defects, wormhole, and lazy S are defects associated with TQF. To limit the occurrence of these defects, hybrid TQFs are used. These include the use of secondary energy sources such as the laser beam, the inductive coil. However, these secondary energy sources have demerits, which lead to weld defects. The ultrasonic approach is used because it promotes material flow, which increases the rate of penetration, while the increase in peak temperature is negligible for large welded joints. Ultrasonically assisted FSW also contributes to continuous frequency oscillation, resulting in high-quality welded joints. Ultrasonic vibrations allowed the formation of very fine grains in the SZ thus increasing its hardness and strength compared to other regions. Fine grain sizes increase the grain boundaries hindering the dislocation movement, so causing an increase in the tensile strength and hardness.

Table 4.4 Mechanical properties of magnesium alloy friction stir welding (FSW) joints [37]

Alloys	Preparation technology	Tensile strength (MPa)	Elongation (%)
AZ31	Extrusion sheet	251	13.2
	FSW	231	9.4
AZ61	Extrusion sheet	308	15.2
	FSW	269	9.6
AZ91	Casting sheet	114	–
	FSW	114	–
AM60	Thixomokling	210	5
	FSW	195	5

4.4.3 Friction Stir Welding of Titanium Alloy

FSW of titanium alloys is mainly applied to the aviation industry, oil pipe-lines, and offshore platforms. Experimental results indicated high welding speed and excessive welding pressure would influence the welding quality, which was consistent with the welding of aluminum or magnesium alloys. Although the titanium alloy possessed a high melting point and high-temperature strength, the process specification of FSW of titanium alloy was relatively small compared with other non-ferrous metals. Due to severe plastic thermo-mechanical deformation the grains are refined in the WN. In the TMAZ/HAZ, deformation, and continuous dynamic recrystallization (CDRX) associated with higher temperature are in competition to define the microstructure in this region. Due to complex deformation, and the irregularity observed in the temperature profiles of the FSW of titanium alloys, the evolving microstructure, and the textures persist as the material flows around the tool and deposits in its wake. The presence of simple shear crystallographic texture as it evolves in the direction of tool rotation can be observed due to bending, and breaking of the lamellar microstructure of Ti-6Al-4 V. The transformation, nucleation, and growth of the respective phases, also depends on the amount of α and β phases present in the base material. The grain refinement of near α, and α + β alloys has been investigated by several researchers. Ti-6Al-4 V, by far, is the most studied titanium alloy (commonly known as the workhorse of the industry) for friction stir welding. Edwards and Ramulu [39] have studied FSWed 12 mm thick Ti-6Al-4 V sheets with a water-cooled W-La alloy with 25 mm shoulder diameter and a tapered pin of 6 mm in length, adopting different rotation and traverse speed to study the peak temperatures and the resulting microstructure in the weld nugget. It is found that higher rotation speed and lower traverse speed resulted in a coarser grained microstructure due to increased temperatures and extended exposure times. The base material with an α/β lamellar microstructure evolved into a fully lamellar microstructure with refined prior β grains.

One of the most agreed mechanisms for the grain refinement in the TMAZ is considered to be CDRX related to shear-induced lattice rotation. According to the literature, the root of the CDRX in FSW can be found in the very early studies on FSW of aluminum alloy. However, some recent works mentioned the discontinuous DRX. In the first place, although sufficiently high strain rate can be predicted, it is difficult to prove DRX in the FSW process where the plastic deformation stage and post heat affected stage (cooling stage) cannot be separated completely. Therefore, further research in a controlled environment that accounts for stage by stage observation of microstructure needs to be carried out in order to understand the true mechanism behind. In case of FSW of titanium alloys where material in the WN (and around) is constantly compressed (from tool shoulder and backing anvil), and sheared (due to tool rotation and traverse), imparted plastic deformation and resultant temperatures play a dramatic role in the development of microstructure. The assumption of simple mechanism either CDRX or DRX seems reasonable if heating and cooling are the only two factors governing the evolution of microstructure. In addition to these two factors, relaxation of material under compressive and shear forces as the tool departs from one location to another, followed by the impact of heating and cooling rate on β transus temperature of titanium alloys, are few other dominating parameters in thorough understanding of microstructural evolution in FSW of titanium alloys. Introduction of other parameters (i.e. β transus temperature and the extent of phase transformation mechanism $\beta \rightarrow \beta + \alpha$) will add more to the complexity of FSW. Zhang et al. studied the effect of rotational speed on the microstructural evolution in the HAZ [40]. The HAZ contained a bimodal microstructure consisting of primary α and α/β lamellar microstructure within prior β grains. The prior grain size increases with rotational speed. As the rotational speed increases, higher temperatures associated with the processing result in the CDRX, with the evolution of rather bigger prior β grains in the HAZ. According to Wu et al. [41], the deformation mechanism in the thermos-mechanically affected zone (if divided into TMAZ 1, beside BM, and TMAZ II beside WN) is governed by dynamic recovery (DRV) in TMAZ I and by continuous dynamic recrystallization (CDRX) in TMAZ II. A similar pattern has also been observed by Pilchak et al. [42] in an investment casted FSWed Ti-6Al-4 V sheet.

The joint performances of FSWed titanium alloy is closely related to welding process and post-welding process. Zhang et al. [40] evaluated the mechanical properties of FSW welded joints. They reported that the mean hardness distribution values in the mixing zone (SZ) decrease with increasing tool speed. The hardness of the joints in the heat affected zone is the lowest, and the hardness decreases the least with the increase of tool speed. The tensile test results show that, compared with the base material, the strength and ductility of the welded specimens are very low, and the mechanical properties of the weld decreases with the increase of the speed of the FSW tool. Ramulu et al. [43] studied the tensile properties of welds under different conditions during welding, welding + stress relief and welding + superplastic forming. The welded sample has higher yield strength and tensile strength than the base material. However, the elongation of the weld is lower than that of the base material. After stress relief treatment, the strength of welded joint is improved slightly and the elongation remains unchanged. Liu et al. [44] evaluated the mechanical properties of welds prepared at constant rotation speed and different transverse velocities. The hardness of weld is lower than that of matrix under all conditions, and the hardness of stirring zone is the lowest. The hardness of mixing zone increases with the increase of welding speed. The agitation zone involves dynamic recrystallization encouraged by frictional heat generated by tool rotation and by plastic deformation performed by the pin tool. Liu et al. [44] also reported that compared with the base material, the tensile strength and ductility of the weld were lower, and the mechanical properties increased with the increase of welding speed. Tensile properties showed that all joints failed in areas with low weld hardness. It is known that the strength of a metal material is proportional to its microhardness value, so a reduction in joint strength can be described by a reduction in the microhardness of the weld. With the increase of welding speed, the recrystallization softening degree of SZ zone decreases, and the mechanical properties of weld are improved under the influence of welding speed. Zhou et al. [45] reported that the weld zone had a minimum hardness value on the base metal and began to decrease with the increase of tool speed. The higher the tool speed, the lower the hardness, because the higher heat input helps the microstructure coarsening. They reported that the transverse tensile properties showed that the elongation and strength of the weld was lower than that of the base material. The tensile stress of weld decreases with the increase of tool speed.

4.4.4 Friction Stir Welding of Steel

The FSW of ordinary low carbon steel is relatively easy, and good quality joints can usually be obtained, and the joint structure is mainly ferrite with good toughness [46, 47]. With the increase of C or alloying element content, the increase of C equivalent makes the hardenability of steel tend to increase, and the parameter selection of FSW is particularly important. At first, most of the FSW studies of steel used high heat input parameters, and the peak temperature of the weld core area could reach 900–1200 °C, which was higher than the complete austenitizing temperature A3, and the hard and brittle phase transition structures such as bainite and martensite were usually obtained during the cooling process [48]. Cui et al. [49] used low heat input parameters (200 r/min, 400 mm/min) to control the peak temperature of FSW below the austenitic phase transition temperature A1, and obtained the mixed structure of ferrite and Fe3C particles in S70C (0.72%C, mass fraction) high-carbon steel joint, while the coarse hard brittle martensite structure was generated under high heat input parameters.

Chung et al. [50] conducted impact toughness tests on the structure of the welded core area showed that compared with the impact absorption energy (24.20 N mm) under high heat input (> A3), the impact absorption energy (24.20 N mm) under low heat input (The impact absorption energy of the < A1) weld core area was significantly increased (67.62 N mm). With the gradual promotion of lightweight, the application of low-carbon low-alloy high-strength steel is increasing, so the attention to its welding is also increasing. Generally, high-strength steels have ultra-fine grains, high-density dislocations and substructures, and most of them are in a highly metastable state, which is easily transformed under the influence of heat input, which brings great difficulties to welding. Therefore, for high-strength steels, especially ultra-high-strength steels (> 1 GPa), although the heat input has been significantly reduced compared with melt welding, the softening of HAZ under conventional parameters is still very obvious [31]. Ghosh et al. [51] found in the FSW study of M190 martensitic steel that the hardness of HAZ is even lower than 50% of the base metal. More seriously, hard and brittle martensitic structure is easy to be generated in the welded core area under conventional parameters, thereby reducing the mechanical properties of the joint. Obviously, the selection of low speed and auxiliary cooling to optimize the structure of the weld core area and reduce HAZ softening are the research focuses of high-strength steel FSW. Xue et al. [52] use auxiliary water cooling to control the peak temperature of FSW in the two-phase region (A1–A3), so that the duplex structure of ultra-fine ferrite and martensite is obtained in the weld core area, which can become the ideal weld structure of high-strength steel FSW. The superfine bi-phase microstructure of the welding core zone has excellent mechanical properties, and the uniform elongation of 7% is still obtained when the yield strength is as high as 1.1 GPa. In addition, by using low-heat input FSW process to control the peak temperature of the welding core zone below A1, ultrafine ferrite microstructure can be obtained, grain size can be refined to 300 nm, and mechanical properties of the joint can also be improved [53].

Most of the FSW research on stainless steel focuses on single-phase austenitic stainless steels, such as 304, 316 and high nitrogen stainless steels. For such stainless steels, the delamination energy is low, and its high-temperature thermal deformation mechanism and recrystallization behavior are more complex than those of aluminum alloys [54]. Nandan et al. [55] showed that the viscoplastic deformation and temperature field simulation in the FSW process of 304 stainless steel showed that a violent plastic flow occurred around the stirring needle, the main heat transfer mode was convection conduction, and the overall temperature field was asymmetrically distributed along the welding tool, which would have a huge impact on the recrystallization behavior of the joint. It is found that under the strong plastic deformation and heat of FSW, the dynamic recrystallization of stainless steel occurs at the beginning. However, due to the inhomogeneity of deformation, the overall dislocation density in the weld shows an uneven distribution state, and these areas with higher dislocation density continue to undergo static recrystallization during the subsequent shaft shoulder heating process, thus showing different orientation characteristics [56]. The study of high nitrogen stainless steel by Du et al. [57] shows that although the structure of the welded core area has been significantly refined, there are not only scale differences between the upper and lower parts of the welded core area. At the same time, the ratio of characteristic grain boundaries varies greatly. A certain amount of ferrite is usually retained in the joint structure of stainless steel fusion welding. Although the temperature of FSW process is lower than that of fusion welding, abnormal phase transformation will occur under the action of severe plastic deformation, leading to the formation of ferrite. And residual and large deformation accompanied by recrystallization of ferrite process will be under certain welding parameter to accelerate the precipitation of phase. Especially under the condition of high heat input, the welding nuclear area at the bottom generates a lot of phase deposition, which seriously deteriorates the corrosion resistance of joint.

In recent years, apart from single-phase austenitic stainless steel, many researchers have also conducted FSW studies on ferritic stainless steel and biphase stainless steel. For ferritic stainless steel, the core zone of FSW is still ferritic, the grain is obviously refined, and the proportion of small corner boundary is increased. Studies on duplex stainless steel show that the ratio of duplex structure in the core zone is similar to that of the base material. FSW shows obvious advantages over fusion welding in the welding of stainless steel. For 304 stainless steel, the welding speed can be as high as 1200 mm/min, and the welding quality is excellent. In the tensile test, the joint breaks on the base metal. For ferritic and duplex stainless steel, the joint of FSW

Table 4.5 Mechanical properties of friction stir welding of steel [37]

Steel	Yield strength (welds/parent metal) (MPa)	Tensile strength (welds/parent metal) (MPa)
C–Mn steel	1040/1400	1230/1710
HSLA-65	597/605	788/673
590 duplex stainless steel	496/340	710/590
304L	51/55	95/98
316L	434/388	641/674
2507 Super Duplex	762/705	845/886
201	193/103	448/406
600	374/263	719/631

is also often broken in the base material, and can be bent to 180° without fracture [58]. Corrosion resistance is a very important property of stainless steel. The low heat input of FSW makes the corrosion resistance of stainless steel joints better than the solidification structure formed by traditional fusion welding, especially for duplex stainless steel. Generally, the cavitation corrosion resistance of dual-phase stainless steel joints is significantly improved compared with the base material, and the passivation performance is gradually improved with the refinement of grain. Li et al. [32] found that the corrosion resistance of high-nitrogen stainless steel joints is very close to that of the base material. The research results of Zhang et al. [40] showed that the pitting potential of HAZ did not decrease significantly, but the pitting resistance of the welding core area was better.

The mechanical properties of friction welded joints in steel materials are shown in Table 4.5 [37]. The tensile strength of FSW joints in some materials is lower than that of the parent metal, such as C-Mn steel. And the mechanical properties of friction welded joints of some materials are close to or even higher than the parent metal, such as 590 duplex stainless steel. This phenomenon is related to the grain composition of the parent metal and its sensitivity to grain size. For example, the percentage of ferrite in a duplex stainless steel FSW joint is similar to that of the parent metal, and the grain size of the weld is greatly reduced. As a result, the performance of welded joints is improved significantly.

4.4.5 Friction Stir Welding of Dissimilar Materials

With the rapid progress of science and technology, the application of new materials and new processes is becoming more and more extensive, and the requirements for structural design are becoming more and more demanding. Generally, during the use of the whole component, the load and external environment of each part are not consistent, and the requirements for material properties are not the same. The application of dissimilar material welding structure can maximize the use of the

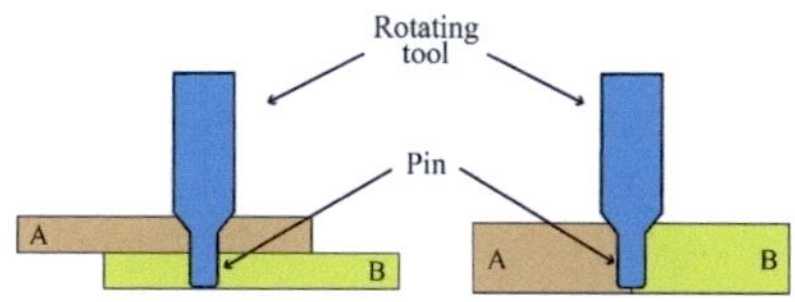

Fig. 4.8 Two main forms of friction stir welding of dissimilar materials

respective advantages of the material, so it has received more and more attention in recent years [33]. Due to the large difference in physical and chemical properties between dissimilar materials, it is not suitable to use fusion welding, so flash welding, friction welding and FSW and other solid phase joining processes have become feasible welding methods for dissimilar materials. Compared with other methods, FSW process is simple, energy saving, high efficiency, and has strong applicability, with wider application prospects. The two main forms of friction stir welding of dissimilar materials are shown in Fig. 4.8. In this section, the following four typical dissimilar material systems are selected for discussion based on the interfacial reaction type.

(1) System with strong interfacial reaction

The typical representative of such systems is Al-Mg dissimilar metal system, Al and Mg are more active elements, although the melting point, thermal conductivity and other physical properties are not much different, but there are differences in crystal structure, and it is easy to react between the two to form various brittle intermetallic compounds (IMC), and even local melting phenomenon, resulting in a sharp decline in the mechanical properties of the joint. Therefore, in recent years, scholars from various countries have made many efforts on how to inhibit the formation of Brittleness IMC in Al-Mg connectors, and some have achieved certain results. It can be seen from the Al-Mg binary phase diagram [59] that at a lower temperature (437 and 450 °C) two kinds of eutectic reactions can occur between Al-Mg, which will lead to the formation of a large number of brittle phases, and the occurrence of liquid phases is very easy to produce solidification cracks. In the selection of FSW parameters, most researchers use lower heat input parameters to avoid or reduce the generation of eutectic reactions, so the range of FSW parameters with better forming quality is generally narrow. Fu et al. [34] obtained the process window of 6061 aluminum alloy-AZ31 magnesium alloy dissimilar metal FSW weld forming, and at low tool speed (600–800 r/min) and low welding speed (30–60 mm/min), defect-free joints can be obtained, and the parameter selection is not easy to appear holes, tunnels, cracks and other defects. In addition, the offset of the stirring pin and the relative position of the welded material also have obvious effects on the forming and mechanical properties of the joint, but there is still no consensus on which material is located on the forward side and how the stirring pin is most conducive to joint forming. When the Al-Mg joint has no macro or microscopic defects, the amount and distribution of IMC generation determine the mechanical properties of the joint. Most studies have shown that the Al-Mg linker interface mainly generates two IMCs, $Al_{12}Mg_{17}$ and Al_3Mg_2 [60]. In order to reduce the amount of IMC generated, some researchers use auxiliary cooling to reduce the welding temperature. The study of Mofid et al. [35] showed that the application of water and liquid nitrogen cooling during the FSW process could reduce the welding temperature by 46 and 53 °C, respectively. However, even in the cooling mode, there is still too much IMC generation, and the mechanical properties of the joint are not significantly improved. In addition to applying cooling, researchers have inhibited or attenuated the reaction between Al-Mg by adding third-party elements.

(2) Medium interface reaction system

Such systems are mainly represented by aluminum alloy and high melting point metal welding, such as Al-Fe, Al-Cu and Al-Ti and other systems. The melting point, thermal conductivity, hardness and other physical and mechanical properties of dissimilar metals of such systems vary greatly, and IMC will also be generated during FSW, but proper parameter control can greatly reduce its generation, and less IMC is conducive to improving the bonding strength. This section will discuss the FSW of the Al-Cu system as an example. Similar to FSW of other dissimilar metals, in addition to the conventional speed, welding speed and other parameters, the offset of the stirring pin and the relative position of the welded material have obvious effects on Al-Cu welding. At first, in order to reduce the wear of the stirring needle, the stirring needle is often biased to the softer Al side, and later more and more research results show that the stirring needle bias is conducive to weld forming and improving the welding quality. The vast majority of studies show that the best weld quality can be obtained when most of the stirring needle is located on the Al side. In addition, when the harder Cu is located on the forward side, it is conducive to the formation of high-quality welds, otherwise it is prone to tunnel defects, which is more serious in the stirring pin bias mode. This is due to the fact that in the weld forming area at the rear of the stirring pin, the material flows from the receding side to the forward side, and if the harder material is located on the receding side, it cannot flow smoothly to the forward side, thus forming a tunnel defect [61]. In order to avoid excessive IMC generation, better weld quality can often be obtained by using lower heat input, such as Xue et al. [62] at a low speed of 600 r/min to obtain a defect-free joint. There are many particles of different sizes distributed in the Al matrix in the welding core area, the larger size is Cu particles, and a thin IMC layer is generated around it, so as to achieve good metallurgical bonding with the Al matrix, while the smaller Cu particles completely react with the Al matrix to generate IMCs such as Al_4Cu_9, Al_2Cu and Al-Cu. There is an obvious interface between the Al substrate and Cu in the weld core area, and scanning electron microscopy (SEM) shows that the thickness of the IMC layer here is very thin, only about 1 μm. Transmission electron microscopy (TEM) analysis showed that the interface layer showed a distinct three-layer structure, with a discontinuous α-Al solid solution layer near the Al side, and an Al_4Cu_9 layer near the Cu side, which contained a small amount of Al-Cu phase and Al_2Cu layer in the middle. It is precisely due to the appropriate amount of IMC generation that a good metallurgical bond is achieved between Al-Cu, which breaks in the heat affected zone (HAZ) of the Al side during tensile, and the strength coefficient reaches 90%, and it can be bent to 180° without fracture. However, when the heat input is high, the thickness of the IMC layer at the interface increases significantly, and a mixed structure composed of IMC and Cu layers appears, and the mechanical properties of the joint decrease significantly. Obviously, how to optimize the FSW process so that the interface generates a uniform and continuous IMC thin layer is the key to achieve the high-performance connection of such system alloys.

(3) No interface reaction system

The typical representative of such a system is the Mg-steel system, the melting point of Fe is much higher than that of Mg, and even exceeds the boiling point of Mg, and the crystal structure of the two is different, and there is almost no solid solution between Mg and Fe and IMC can not be generated, and the wettability is very poor, which brings great difficulties to its welding. Present, the research on Mg-steel FSW is still relatively limited, and mainly based on spot welding and lap welding are the mainstay [63]. Many studies have shown that the FSW joint between magnesium alloy and Zn-plated steel plate is better than that of joints with non-plated Zn steel plate, because the Zn plating layer reacts with the Mg matrix to form the eutectic structure of Mg-Zn, which increases the wettability of Mg on the steel surface. Zhang et al. [64] used a retractable stirring needle to achieve high-quality FSSW of Mg-steel, and the tensile shear strength of the joint was up to 8.7 kN. It is found that when the lap form of Mg is located in the upper part, the stirring needle has less wear, but due to the low welding temperature, the mechanical bonding between the Mg-steel interface is mainly shown. When Mg is located in the lower part, the interface reaction temperature is increased, the precipitation of Al elements from the Mg matrix is promoted, and a variety of complex compounds such as Mg-Zn and Al-Fe are generated at the Mg-steel interface, forming a metallurgical interface and significantly increasing the joint strength. Therefore, the Zn plating layer only plays the role of promoting interface wetting, and the metallurgical bond between Mg-steel is inseparable from the role of Al elements in the magnesium alloy matrix. In addition, through the special design of the stirring needle, the steel matrix is fully agitated to produce more flash structure, which is nailed to the Mg matrix, which can greatly improve the mechanical bonding effect of the joint, thereby improving the performance of the joint [63].

(4) Metal-metallic glass and metal-plastic systems

In recent years, many researchers have explored FSW for new dissimilar material systems, such as metal-amorphous systems and metal-plastic systems. Wang et al. [65] successfully obtained a defect-free joint of 7075 aluminum alloy and zirconium-based amorphous by using the process of biasing the stirring needle to the Al side, the results show that the welding core region contains a large number of uncrystallized amorphous fragments, no obvious reaction is found between amorphous and Al, and the joint strength can reach 74% of the aluminum alloy base metal. Sun et al. [66] conducted FSW on pure Cu and zirconium-based amorphous by two methods: butt welding and spot welding, In those studies, good mechanical properties were obtained, and in addition to the amorphous fragments in the welded core area, a mixed structure composed of amorphous strips and pure Cu was also found at the interface. Obviously, FSW can realize high-quality welding of metal and amorphous, but the specific interface bonding mechanism is still unclear. The physical and chemical properties between metals and plastics are completely different, so how to achieve the combination between the two when FSW is the key. The generation of heat in the FSW process will cause the plastic to produce bubbles, and the presence of a small amount of bubbles at the interface can generate pressure to promote the welding of metal and plastic, but excessive bubbles will significantly reduce the joint performance. Therefore, how to control the amount of bubbles produced is a problem that must be paid attention to in the FSW process. Liu et al. [67] showed that excellent mechanical properties can be obtained when the amount of bubbles is less than 8%. For the FSW of aluminum alloy and plastic, the tensile shear strength of the joint is as high as 43 MPa when the Al layer is present on the alloy surface, which is significantly higher than that of the joint without Al (28 MPa) [68]. In addition, plasma electrolytic oxidation (PEO) treatment on the metal surface can form a combination of PEO coating and polyethylene on the surface of the magnesium alloy. At the same time, polyethylene extends into the coating to form a micromechanical interlocking effect, which also greatly improves the adhesion between magnesium alloy and polyethylene joint.

FSW has shown good application prospects in the welding of dissimilar materials, and is expected to become an ideal process for dissimilar material joining, but it still needs to be studied in terms of process optimization and connection mechanism. For violent interfacial reaction systems, how to avoid interface overreaction by improving FSW parameters and introducing third-party elements is the key to future research. For medium interface reaction systems, adjusting the FSW parameters to generate uniform and continuous IMC thin layers at the interface is the key to obtaining excellent joint performance, and its fine interface microstructure still needs to be studied in depth. For interface non-reactive systems, how to improve interface wettability and give full play to the role of alloying elements such as Al and Zn to promote the combination of interface metallurgy is still a problem to be solved. For new systems such as metal-amorphous and metal-plastic, while expanding the research system, the in-depth study of the interface bonding mechanism will be the focus of attention.

4.5 Applications of Friction Stir Welding

4.5.1 Application of Friction Stir Welding in Aerospace

Friction stir welding has a wide range of applications in the aerospace industry. It covers the fields of aircraft structure manufacturing, spacecraft manufacturing, gas turbine engine manufacturing and lightweight design. The technology can reduce welding deformation and material damage, and improve the safety, reliability and performance of aircraft [69]. The application of FSW to fabricate launch vehicle tanks is shown in Fig. 4.9. Its application is mainly reflected in the following aspects.

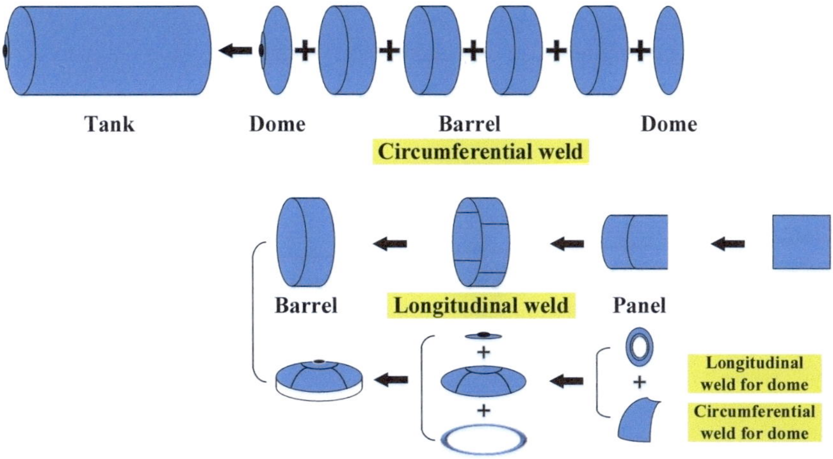

Fig. 4.9 Schematic diagram for friction stir welding of launch vehicle fuel tank

Joining of structural parts of aircraft: Structural parts of aircraft usually need to be characterized by high strength and light weight. Friction stir welding can achieve non-melt welding, avoiding the melting and solidification process in traditional welding methods, thus reducing welding deformation and the heat-affected zone of the material. This type of welding can effectively maintain the original properties of structural components and improve the quality and strength of the connection. For example, friction stir welding is used to join structural parts such as wings, fuselages and tails of aircraft, providing uniform, high-strength joints.

Connection of engine parts: Aero-engine is the core component of the aircraft, which has very high requirements for connection joints. Friction stir welding can achieve high-quality welded joints that maintain the original properties of the material and avoid defects such as porosity and cracks in traditional welding methods. For example, friction stir welding is used to join components such as blades, turbine discs and combustion chambers of aero-engines, providing a reliable connection that improves engine performance and life.

Manufacture of fuel tanks: Fuel tanks are an important part of aircrafts, which require high sealing and reliability of connecting joints. Friction Stir Welding can achieve defect-free, high-strength connecting joints to ensure the hermeticity and safety of fuel tanks. For example, friction stir welding has been used to join fuel tank wall and floor plates in aircraft, providing a reliable connection that ensures the safe storage and supply of fuel.

The Eclipse Company successful used FSW technology in the manufacturing of Eclipse N500 type business aircraft for the first time in large-scale, including aircraft skin, wing help, string support, aircraft floor and structural components. It not only greatly improves the quality of the connection, but also makes the production efficiency increased by nearly 10 times faster than automatic riveting, 6 times faster than manual riveting, and 60 times faster than the total cost savings. The use of friction welding instead of fasteners simplifies the structural design of the floor and improves the efficiency of the production of the components, production costs reduced by 20%.

4.5.2 *Application of Friction Stir Welding in Automobile Manufacturing*

Friction stir welding is widely used in the automobile manufacturing industry, covering areas such as body building, engine manufacturing, wheel manufacturing and lightweight design. It can provide high-strength, high-quality welded connections, while reducing deformation and porosity, to meet the requirements of structural strength, lightweight and durability in automobile manufacturing, and help to improve the performance, safety and fuel economy of the car. Schematic diagram and actual structure of fillet joint FSW for main battery frame of a vehicle are shown in Fig. 4.10 [70]. Its application is mainly reflected in the following aspects:

Bodywork: Friction Stir Welding is used to join the bodywork of automobiles, e.g. joining body panels, doors, roofs, etc. This welding method provides a high-strength joint with reduced weld distortion and porosity. Friction stir welding can also be used to join different types of materials such as aluminum alloys and steel.

Engine components: Friction stir welding can be used to join engine components such as crankcases and cylinder heads. This welding method provides a high-strength connection and good durability in high-temperature and high-pressure environments. Friction Stir Welding also reduces welding-induced stress concentrations and distortion, improving engine performance and reliability.

Automotive chassis: Friction stir welding can be used to join automotive chassis components such as suspension and steering systems. This welding method provides

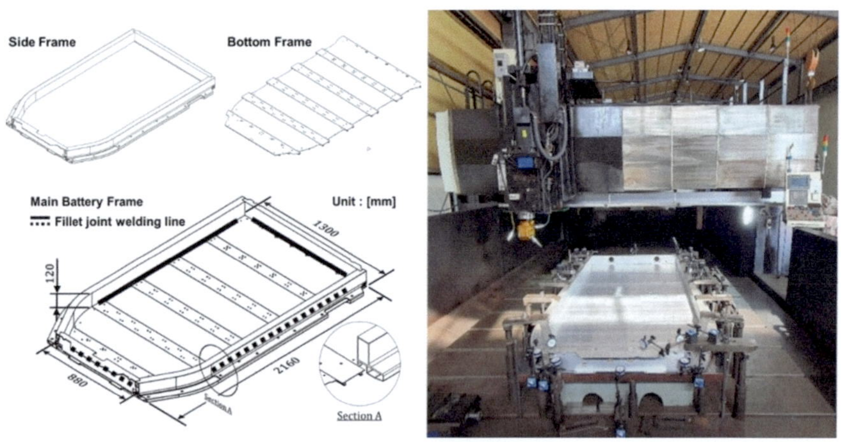

Fig. 4.10 Schematic diagram and actual structure of fillet joint FSW for main battery frame [70]

strong connections and can withstand complex dynamic loads. Friction Stir Welding can also increase the rigidity and durability of the chassis, improving the handling and safety of the vehicle.

Lightweight construction: Friction Stir Welding is widely used in the manufacture of lightweight structures for automobiles. By using lightweight materials (e.g. aluminum alloys) and friction welding technology, the weight of a vehicle can be significantly reduced, resulting in improved fuel efficiency and reduced exhaust emissions. For example, Tesla Motors uses a large number of aluminum alloy friction welded joints in its electric models to achieve higher energy efficiency and driving range.

4.5.3 Application of Friction Stir Welding in Marine Ships

Friction stir welding has important application value in ship manufacturing. It covers the fields of ship hull manufacturing, ship equipment manufacturing, ship pipeline manufacturing and lightweight design. It provides high-strength, high-quality welded joints, which can improve the welding quality, reduce welding defects, and improve the lightweight design of the ship, the strength of the connection and environmental performance, and has become one of the indispensable welding technologies to promote the development of the ship manufacturing industry. The FSW process of mild steel plates used in shipbuilding is shown in Fig. 4.11 [71]. Its application is mainly reflected in the following aspects:

Hull Structures: Friction Stir Welding is used to join hull structures such as plates, bottoms, and holds. This welding method can provide high strength connections while reducing welding-induced distortion and defects. Friction Stir Welding can also be used to join different types of materials, such as aluminum alloys and steel, to meet the needs of different parts of shipbuilding.

Ship decks and deck equipment: Friction Stir Welding can be used to join ship decks and deck equipment, such as ship cranes and ship deck coverings. This welding method provides a strong connection and can withstand the vibrations and loads of

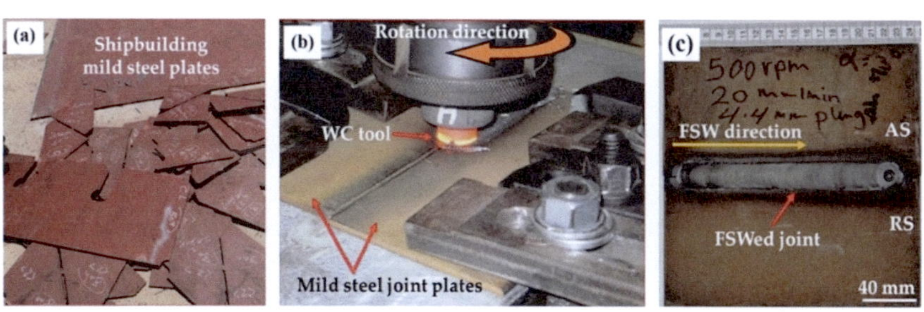

Fig. 4.11 **a** Mild steel shipbuilding plates; **b** The FSW process of mild steel plates; **c** a top view of the produced butt joint [71]

a ship in the complex environment of the sea. Friction Stir Welding can also reduce welding-induced stress concentrations and fatigue cracks, improving the reliability and durability of ships.

Ship pipework: Friction Stir Welding can be used to join ship pipework such as fuel and water pipes. This welding method provides a sealed connection and can withstand high pressure and high temperature environments. Friction Stir Welding also improves the safety and reliability of marine piping systems by reducing porosity and air leakage problems caused by welding.

Lightweight construction: Friction Stir Welding is used to create lightweight ship structures. By using lightweight materials (e.g. aluminum alloys) and Friction Stir Welding technology, the weight of ships can be significantly reduced, resulting in improved fuel efficiency and reduced emissions. For example, a number of high-speed passenger ships and military vessels have used Friction Stir Welding to join hull structures and deck equipment to improve performance and operational efficiency.

4.6 Other Friction Stir Processing Methods

4.6.1 Friction Stir Spot Welding

Friction stir spot welding (FSSW) is the modified version of the FSW technique originally developed by Mazda Corporation, Japan, in 1993 for automotive applications [72]. FSSW process was developed originally to join thin sheet aluminum alloys within a short period. FSSW is a solid-state welding process which uses a non-consumable tool with no traverse movement [73]. Under high loading conditions and frictional heat, the material gets plastically deformed. The workpieces get joined under the forging action of the tool onto the workpiece [74]. FSSW is a transient process as it takes place for a very short cycle time, and is used for joining thin metal sheets with minimum material loss [75]. Some of the advantages of the process include minimal distortion, low residual stress, low operational hazards, and fine and uniform weld structure [75, 76]. The disadvantages mainly include the formation of keyhole and corrosion of the keyhole, inflexibility in comparison to the arc welding process as it needs a backing anvil and rigid support and is applicable to thin sheets only [76, 77].

FSSW process works with three stages as follows:

(1) **Plunging**: Non-consumable tools can move into the work material in a controlled manner. At the interface, frictional heat is generated that could soften the tool material. The applied downward force with rotation ensures the material plasticizes and offers strong joint action.
(2) **Stirring**: After reaching a certain depth (desired) called plunge depth, the tool stays at that point for a desired amount of time, called the dwell time.

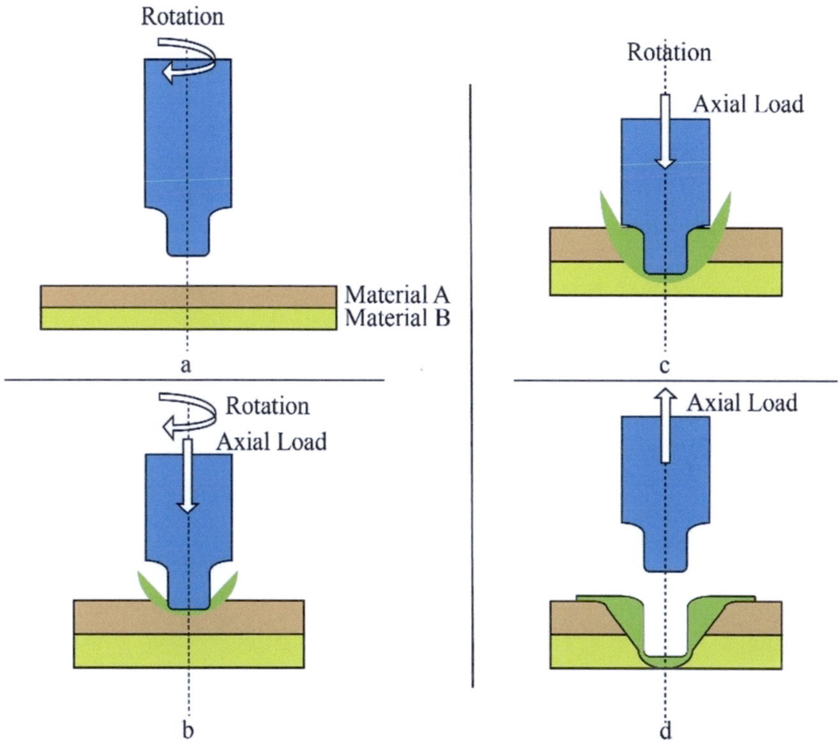

Fig. 4.12 Schematics of FSSW: **a** Tool rotation, **b** plunge, **c** dwell (stirring action), **d** retraction

(3) **Drawing Out**: After the dwell time is passed, the tool is taken out. This leaves behind a keyhole or shape of the tool on the workpiece [75]. The schematic representation of the FSSW process is presented in Fig. 4.12.

The joint partition of FSSW can be divided into four zones: plastic zone, dynamic resting layer, HAZ, and base material. The plastic zone is the area where both sides of the "keyhole" are directly affected by the shoulder and the stirring needle, and a large plastic extrusion deformation occurs in this area under the heat and force of the shoulder and the stirring needle; the dynamic resting layer is the area where the material at the bottom of the keyhole moves very slowly when it is extruded by the stirring needle [78]. Extrusion, plastic deformation experienced by the larger: cooked impact zone refers to the plastic zone and the transition zone of the parent material, this region in the shoulder and the stirring needle thermal action of the larger plastic extrusion deformation. Among them, the plastic zone and dynamic resting layer is the dynamic recrystallization zone formed by the flow of plastic material by the rotation of the stirring needle and the shaft shoulder, which is a region of fine grains and high hardness. FSSW technology was used by Mazda in 2003 for the production of the hood and rear door of the sports model Mazda Rx-8.

4.6.2 Friction Stir Lap Welding

FSLW is derived from FSW (**FSLW**), wherein the workpieces are overlapped to a certain distance [79]. The overlapped region is welded viz. non-consumable tool. The tool rotates at a certain rpm and traverses over the lap joint. Due to the friction on the surface, a large amount of heat is generated at the interface. The material gets plastically deformed, and the material is softened locally. FSLW is conducted by conducting experiments with the appropriate setting of machine parameters. If the desired height is achieved, the process is stopped; else, repeat the process till the part ensures the desired height. FSLW has various variety of applications in welding steel, copper, titanium, magnesium, and composite metal matrix. In FSLW of two dissimilar materials, the soft sheet is placed on the top and the hard sheet at the bottom. There is a decrease in the effective sheet thickness due to large penetration into the bottom sheet resulting in a hook defect. For example, FSLW of Al-Cu, Al is placed on the top and Cu at the bottom to produce a good lap joint. Figure 4.13 shows the schematics of the FSLW process.

It has been observed that during FSLW of Al-to-Al, Mg-to-Mg, steel-to-steel material flow affects the un-welded lap on the advancing and retreating side. The upward flow of material on the advancing side creates a hook in the un-welded lap. This hook is viewed as a crack, and acts as an origin point for crack growth. High rotational speed or low traverse speed has been found to form a large hook. This eventually reduces the strength of the weld. Detailed quantification of hooking is essential to determine the effect of hooking strength on the weld formed. FSLW experiments are performed for the fixed values of the tool rotation speed of 1600 rpm, resulting in a good mix and uniform distribution of polymer and weld metal formed in the weld zone [80]. The industry trend is shifting from copper to aluminum (Cu is susceptible to segregation and uniform copper distribution particles lead to decreased electrochemical potential and therefore increased corrosion behavior). Still, a complete shift could affect the working system. Aluminum is a potential alternative to solve the existing problem. Welding aluminum and steel by placing aluminum on top and steel at the bottom is another area to explore that ensures potential application for FSLW.

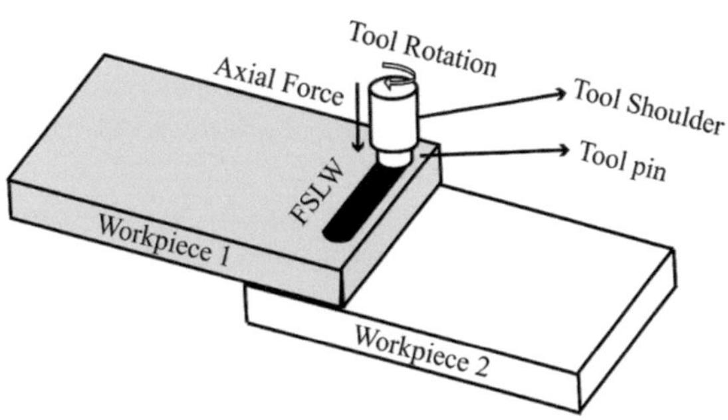

Fig. 4.13 Schematics of the FSLW process

4.6.3 *Stationary Shoulder Friction Stir Welding*

Stationary shoulder Friction stir welding (SFSW) is a new welding method proposed by TWI based on traditional FSW, whose stirring tool consists of a rotating stirring probe and a stationary shoulder that slides on the material surface. During the welding process, the internal stirring needle is rotating, while the external shoulder does not rotate, but only travels in the welding direction, as shown in Fig. 4.14.

It can achieve not only the welding advantages of FSW but also the welding of T-joints and fillet welds. SSFSW is a new derivative welding method based on FSW, which not only has the unique advantages of FSW but also has a series of advantages that cannot be achieved by conventional FSW:

(1) The surface of the weld is smooth and nice and no arc pattern is produced.
(2) Uniform heat input of welding, a uniform organization in the direction of plate thickness, small deformation. No flying edges after welding, smaller thinning in the weld area, improving the effective bearing section of the joint.
(3) The ability to fill the material during the welding process, thus providing the basis for the development of SSFSW additive manufacturing methods.
(4) Welding of T-joints of different thicknesses and complex fillet welds is possible.

Conventional FSW is mainly used for plate butt welded joints, ring butt welded joints, lap joints, and similar joint forms, in the fillet weld the shoulder of the FSW is easy to damage the base material on both sides and the filler material is not easy to ensure. However, Stationary SFSW is designed with a specific shape of the shoulder to exactly match the shape of the fillet weld, to be in close contact with the plate on both sides of the fillet weld during the welding process, and to keep advancing with the stirring tool in the direction of the weld, but not rotating. However, when the cross-section of the static shoulder is a right triangle, the transition angle between the two plates of the obtained fillet weld is right- angled, causing stress concentration and thus affecting the welded joint capacity and fatigue life. To improve the joint form and thus the joint performance, the static shoulder is improved and the fillet

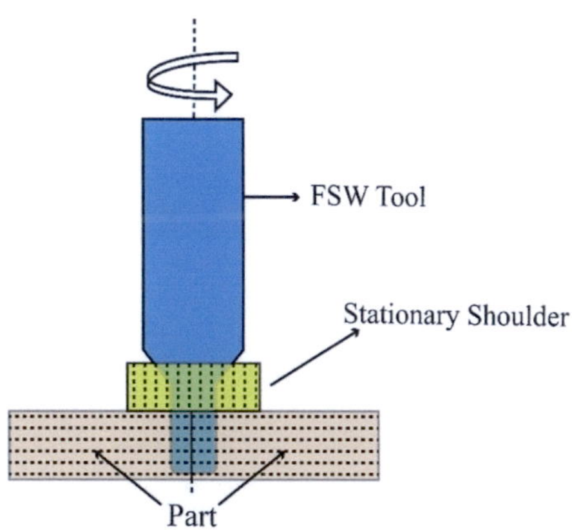

Fig. 4.14 Working principle of stationary shoulder friction stir welding

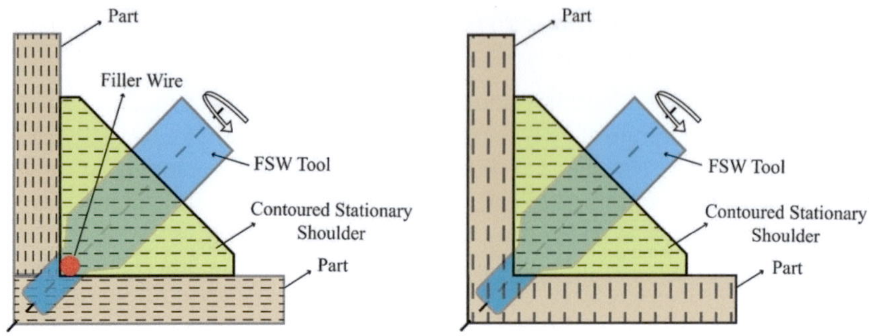

Fig. 4.15 Working principle of wire-filled stationary shoulder stir friction stir welding

weld forming method with fillet transition is realized by the wire filling method, the basic principle and the shape after welding is shown in Fig. 4.15.

4.6.4 Self-reacting Friction Stir Welding

Traditional FSW application in the curved and large structures is to some extent in limitations. Consequently, the self-reacting friction stir welding (SRFSW) technique (also called as bobbin tool friction stir welding) is developed by employing a bottom shoulder to contact the bottom surface of the workpiece. Moreover, the SRFSW tools can allow the distance between the shoulders variable and keep the force constant during the whole welding, which is named as adaptive self-reacting friction stir welding (ASRFSW). SRFSW has not only the clamping end, upper shoulder, and probe, but also a lower shoulder that plays a rigid supporting role. The basic principle of SRFSW is shown in Fig. 4.16. At the beginning of welding, the stirring tool cuts into the base material at a high speed, and the base material reaches a plastic state under the stirring friction of the stirring tool. Compared to conventional FSW, SRFSW has the following advantages:

Fig. 4.16 Basic principle of SRFSW

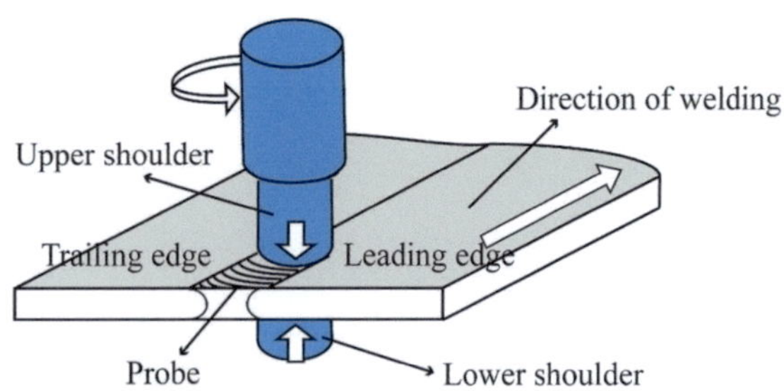

(1) In the SRFSW process, the lower shoulder acts as a rigid backing plate at the back, while heating the workpiece by friction heating, simplifying the tooling and saving costs.
(2) Be better adaptable to different welding conditions, the ability to weld complex shapes and special parts of the workpiece, such as double streamlines, hollow structures, etc.
(3) The needle passes through the upper and lower surfaces of the workpiece during the welding process, effectively eliminating the root defects produced during traditional stir friction welding.
(4) Double shoulder mode to increase heat input, increase welding speed and improve welding efficiency.
(5) SRFSW upper and lower shoulders are heated evenly in the thickness direction of the part to be welded, avoiding the traditional heat input uniform single situation, resulting in uniform heating on both sides and small welding deformation.

4.6.5 Friction Stir Additive Manufacturing

Friction Stir Additive Manufacturing (FSAM) is one of the friction-based additive techniques. In FSAM, each plate is made flat, and dimensions are checked. As the process employs layer-by-layer deposition, all plates should be the same size. This is followed by cleaning the plates and keeping one above the other. A specially designed non-consumable rotating tool with the desired designed shoulder and the pin is injected into overlapping sheets, and the tool is moved along. The friction generated between tool and plates produces intense heat resulting in plastic deformation necessary to join two plates. Figure 4.17 shows the schematics of the FSAM process. Aeroprobe has developed a friction stir additive manufacturing method that realizes the additive manufacturing of the end face filler of the stirring tool, which offers a significant advantage in terms of manufacturing efficiency, material utilization, and process flexibility, as opposed to multi-layer stacked welding methods. FSAM is the tailored version of FSW, showing significant attention in industries (Boeing and Airbus) since 2014. The major advantage of this process is the reduction in material waste, properties better than wrought, structures, and high production rates [81]. The limitations include a higher tool wear rate and inhomogeneous microstructure.

4.6.6 Robotic Friction Stir Welding

Due to the very high flexibility of the robot, which can move in any trajectory in space, the development of robotic FSW systems began abroad in 1997, the sixth year after the invention of FSW technology. After years of research, robotic FSW technology has been applied abroad to the welding of complex structures such as aerospace. In

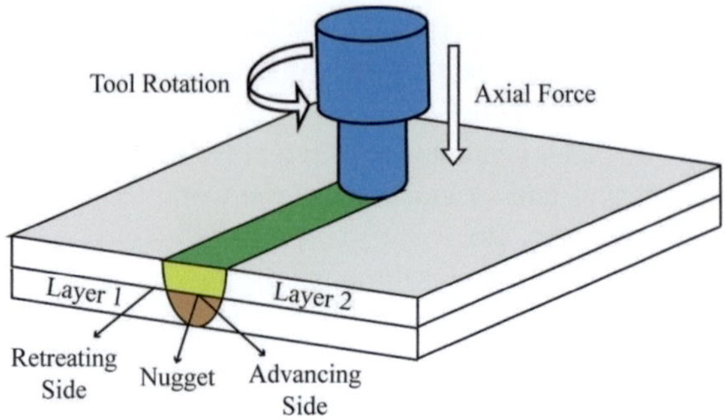

Fig. 4.17 Schematics of FSAM process

order to introduce industrial robots into the manufacture of FSW equipment, well-known companies in various countries have carried out a lot of exploratory work, after more than ten years of research and development, several internationally renowned robotics companies have now launched robot FSW integrated system, such as the United States FSL company based on the ABB robot system developed by the robot FSW system, the robot at the same time integrated pressure control, torque The robot also integrates pressure control, torque control, displacement control and temperature control modules, and develops a user-friendly interface based on the stir friction welding process, and is currently mainly used in research institutes, such as TWI in the United Kingdom and the Beijing Institute of Aeronautical Manufacturing Engineering. Figure 4.18 shows the local morphology of robot friction stir welding equipment and corresponding weld morphology. According to the ABB company reports, the robot friction stir welding has many advantages such as greatly reduced footprint, 50% higher welding speed, high flexibility, one-dimensional, two-dimensional and three-dimensional structure welding, good quality, high-quality welds without flashing, reduced production costs, and no subsequent manual grinding. The friction stir welding solution provided by ABB robots includes two processes: static shoulder and moving shoulder. The motor speed can reach 4000 rpm, and the life of the stirring needle can reach 2000 m. Based on the force control constant pressure control of ABB robot, the maximum downforce can reach 15KN, and the maximum welding speed can reach 3 m/min. The thickness of the welding profile can reach up to 10 mm, the surface of the weld formed under this process is beautiful, and the welded workpiece also has the characteristics of small deformation, stable quality, and smooth corner transition. Robotic friction stir welding can be applied to different fields to provide customers with higher flexible production. Robotic friction stir welding is used in the production of new energy battery box trays, and its weld strength can reach more than 80% of the base metal, and form an excellent weld joint structure. Increase production flexibility and shorten delivery times while maintaining high quality and consistency.

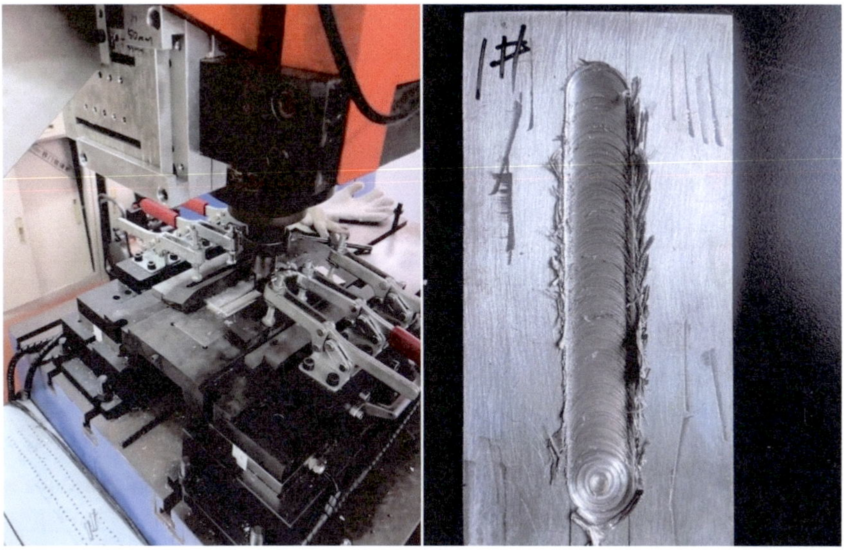

Fig. 4.18 Local morphology of robot friction stir welding equipment and corresponding weld morphology

4.7 Summary and Prospects

Friction stir welding is a welding method that uses high temperatures generated by rotary stirring and heated friction to achieve a material joint. It generates frictional heat by applying axial force and rotational motion between the joining parts, causing local plastic deformation of the materials and forming a uniform bond. This welding method does not require additional filler material and has the advantages of high speed, no porosity, no cracks, high strength, etc. It is suitable for the connection of a variety of metallic and non-metallic materials, and has a wide range of prospects for application, especially in the fields of aerospace, automobile manufacturing, ship construction and rail transport.

With regard to the method of friction stir welding, we can see the following trends and directions of development:

(1) Technical improvement and optimization: as the research on friction stir welding continues, there will be more technical improvement and optimization. For example, improve the design of stirring head to enhance the quality and reliability of joints; optimize the parameter control to achieve more accurate welding process and performance; carry out more material adaptability research to expand the scope of application of Friction Stir Welding and so on.
(2) Expansion of application areas: Friction Stir Welding has unique advantages as a non-molten metal joining method. In the future, it is expected that it will be applied in more fields, such as electric power equipment, petrochemical industry

and so on. At the same time, it may also be extended to emerging areas, such as new energy equipment manufacturing, electronic components packaging.

(3) Welding of new materials: friction stir welding in the traditional welding of metal materials has been well applied, the future will face challenges and opportunities to achieve the welding of new materials. For example, high-strength steel, nickel-based alloys, composite materials. For the special needs of these materials, the need for further research and development of highly adaptable friction welding process.

(4) Automation and Intelligence: With the intelligent development of manufacturing industry, friction stir welding is also expected to achieve automation and intelligence. Through the introduction of machine vision, sensor technology and intelligent control systems, automatic monitoring, parameter adjustment and quality assurance of the welding process can be realized, so as to improve production efficiency and stability.

Overall, as an innovative welding method, friction stir welding will continue to be widely researched and applied in the future. Through continuous technological improvement and expansion of application areas, it is expected to become one of the important joining technologies in the modern manufacturing industry, promoting the development and progress of related industries.

Problems

1. What is friction stir welding? How does it differ from other types of welding?
2. How does FSW work? How does it join two metal plates together?
3. What are the advantages of FSW? In what areas is it widely used?
4. What are the disadvantages of FSW? What are its other limitations?
5. What parameters need to be considered in the FSW process? How do these parameters affect the quality of the weld?
6. What are the effects of tool shape and material selection for FSW on the weld?
7. How is the quality of the weld tested and evaluated in FSW?
8. What are the limitations of FSW compared to other types of welding in terms of its range of applications?
9. What are the future directions for FSW? What new technologies and applications are being developed?
10. What are the applicable materials for FSW? What are their thickness and shape requirements?
11. What defects can occur during the welding process of FSW? How might these defects affect the quality of the weld?
12. What safety issues need to be considered during FSW? How can the safety of the operator be ensured?
13. What are some specific examples of FSW in different applications? What are their success stories?
14. How does FSW compare to other welding technologies in terms of cost? Can it be used in mass production?

15. What are the environmental advantages of FSW? What is its impact on the environment?

References

1. Delijaicov S, Rodrigues M, Farias A et al (2020) Microhardness and residual stress of dissimilar and thick aluminum plates AA7181-T7651 and AA7475-T7351 using bobbin, top, bottom, and double-sided FSW methods. Int J Adv Manuf Technol 108(1):277–287
2. Kumar Rajak D, Pagar DD, Menezes PL et al (2020) Friction-based welding processes: friction welding and friction stir welding. J Adhes Sci Technol 34(24):2613–2637
3. Zhang F, Su XK, Chen ZY et al (2015) Effect of welding parameters on microstructure and mechanical properties of friction stir welded joints of a super high strength Al–Zn–Mg–Cu aluminum alloy. Mater Des 67:483–491
4. Mohanty HK, Mahapatra MM, Kumar P et al (2012) Effect of tool shoulder and pin probe profiles on friction stirred aluminum welds—a comparative study. J Mar Sci Appl 11(2):200–207
5. Bilici MK, Yukler AI (2012) Effects of welding parameters on friction stir spot welding of high density polyethylene sheets. Mater Des 33:545–550
6. Zhao Y, Zhou LL, Wang QZ et al (2014) Defects and tensile properties of 6013 aluminum alloy T-joints by friction stir welding. Mater Des 57:146–155
7. Li B, Shen YF (2012) A feasibility research on friction stir welding of a new-typed lap–butt joint of dissimilar Al alloys. Mater Des 34:725–731
8. Cui L, Yang XQ, Xie YH et al (2013) Process parameter influence on defects and tensile properties of friction stir welded T-joints on AA6061-T4 sheets. Mater Des 51:161–174
9. Heidarzadeh A, Saeid T (2016) A comparative study of microstructure and mechanical properties between friction stir welded single and double phase brass alloys. Mater Sci Eng, A 649:349–358
10. Zhang H, Wu L, Lin SB et al (2004) Friction stir welding of AZ31 magnesium alloy. J Mech Eng 40(8):123–126 (in Chinese)
11. Firouzdor V, Kou S (2010) Al-to-Mg friction stir welding: effect of material position, travel speed, and rotation speed. Metall Mater Trans A 41(11):2914–2935
12. Aissani M, Gachi S, Boubenider F et al (2010) Design and optimization of friction stir welding tool. Mater Manuf Processes 25(11):1199–1205
13. Zhou L, Wang T, Zhou WL et al (2016) Microstructural characteristics and mechanical properties of 7050–T7451 aluminum alloy friction stir-welded joints. J Mater Eng Perform 25(6):2542–2550
14. Ahmed MMZ, Ataya S, El-Sayed Seleman MM et al (2020) Heat input and mechanical properties investigation of friction stir welded AA5083/AA5754 and AA5083/AA7020. Metals 11(1):68
15. Ni Y, Qin DQ, Mao Y et al (2020) Influences of welding parameters on axial force and deformations of micro pinless friction stir welding. Int J Adv Manuf Technol 106(7):3273–3283
16. Razal Rose A, Manisekar K, Balasubramanian V (2011) Effect of axial force on microstructure and tensile properties of friction stir welded AZ61A magnesium alloy. Trans Nonferrous Metals Soc China 21(5):974–984
17. Kumar K, Kailas SV (2008) On the role of axial load and the effect of interface position on the tensile strength of a friction stir welded aluminium alloy. Mater Des 29(4):791–797
18. Aldanondo E, Arruti E, Echeverria A (2017) Friction stir weld lap joint properties in aeronautic aluminium alloys. In: The Minerals, Metals & Materials Series. Springer International Publishing, Cham, pp 109–117

19. Gungor B, Kaluc E, Taban E et al (2014) Mechanical, fatigue and microstructural properties of friction stir welded 5083-H111 and 6082-T651 aluminum alloys. Mater Des (1980–2015) 56:84–90
20. Santos TFA, Hermenegildo TFC, Afonso CRM et al (2010) Fracture toughness of ISO 3183 X80M (API 5L X80) steel friction stir welds. Eng Fract Mech 77(15):2937–2945
21. Arora A, De A, DebRoy T (2011) Toward optimum friction stir welding tool shoulder diameter. Scripta Mater 64(1):9–12
22. Mehta M, Arora A, De A et al (2011) Tool geometry for friction stir welding—optimum shoulder diameter. Metall Mater Trans A 42(9):2716–2722
23. Galvão I, Leal RM, Rodrigues DM et al (2013) Influence of tool shoulder geometry on properties of friction stir welds in thin copper sheets. J Mater Process Technol 213(2):129–135
24. Leal RM, Leitão C, Loureiro A et al (2008) Material flow in heterogeneous friction stir welding of thin aluminium sheets: effect of shoulder geometry. Mater Sci Eng, A 498(1/2):384–391
25. Zhang YN, Cao X, Larose S et al (2012) Review of tools for friction stir welding and processing. Can Metall Q 51(3):250–261
26. El-Batahgy AM, Terad B, Omar A (2013) Effect of friction stir welding parameters on properties of AA6061 aluminum alloy butt welded joints. In: Proceedings of the 1st international joint symposium on joining and welding. Elsevier, Amsterdam, pp 33–40
27. Buffa G, Campanile G, Fratini L et al (2009) Friction stir welding of lap joints: influence of process parameters on the metallurgical and mechanical properties. Mater Sci Eng, A 519(1/2):19–26
28. Motalleb-nejad P, Saeid T, Heidarzadeh A et al (2014) Effect of tool pin profile on microstructure and mechanical properties of friction stir welded AZ31B magnesium alloy. Mater Des 59:221–226
29. Khodaverdizadeh H, Heidarzadeh A, Saeid T (2013) Effect of tool pin profile on microstructure and mechanical properties of friction stir welded pure copper joints. Mater Des 45:265–270
30. Mohan DG, Wu CS (2021) A review on friction stir welding of steels. Chin J Mech Eng 34(1):137
31. Matsushita M, Kitani Y, Ikeda R et al (2011) Development of friction stir welding of high strength steel sheet. Sci Technol Weld Joining 16(2):181–187
32. Li HB, Jiang ZH, Feng H et al (2015) Microstructure, mechanical and corrosion properties of friction stir welded high nitrogen nickel-free austenitic stainless steel. Mater Des 84:291–299
33. Kumar N, Yuan W, Mishra RS (2015) Friction stir welding of dissimilar alloys and materials. Butterworth-Heinemann, Oxford
34. Fu BL, Qin GL, Li F et al (2015) Friction stir welding process of dissimilar metals of 6061–T6 aluminum alloy to AZ31B magnesium alloy. J Mater Process Technol 218:38–47
35. Mofid MA, Abdollah-Zadeh A, Ghaini FM et al (2012) Submerged friction-stir welding (SFSW) underwater and under liquid nitrogen: an improved method to join Al alloys to Mg alloys. Metall Mater Trans A 43(13):5106–5114
36. Threadgill PL, Leonard AJ, Shercliff HR et al (2009) Friction stir welding of aluminium alloys. Int Mater Rev 54(2):49–93
37. Zhang H, Khoshnaw F (2023) Friction stir welding. In: Welding of Metallic Materials. Elsevier, Amsterdam, pp 197–228
38. Klenam DEP, Ogunwande GS, Omotosho T et al (2021) Welding of magnesium and its alloys: an overview of methods and process parameters and their effects on mechanical behaviour and structural integrity of the welds. Manuf Rev 8:29
39. Edwards P, Ramulu M (2010) Peak temperatures during friction stir welding of Ti–6Al–4V. Sci Technol Weld Joining 15(6):468–472
40. Zhang Y, Sato YS, Kokawa H et al (2008) Microstructural characteristics and mechanical properties of Ti–6Al–4V friction stir welds. Mater Sci Eng, A 485(1/2):448–455
41. Wu LH, Wang D, Xiao BL et al (2014) Microstructural evolution of the thermomechanically affected zone in a Ti–6Al–4V friction stir welded joint. Scripta Mater 78(79):17–20
42. Pilchak AL, Williams JC (2011) Microstructure and texture evolution during friction stir processing of fully lamellar Ti-6Al-4V. Metall Mater Trans A 42(3):773–794

43. Ramulu M, Edwards PD, Sanders DG et al (2010) Tensile properties of friction stir welded and friction stir welded-superplastically formed Ti–6Al–4V butt joints. Mater Des 31(6):3056–3061
44. Liu HJ, Zhou L, Liu QW (2010) Microstructural characteristics and mechanical properties of friction stir welded joints of Ti–6Al–4V titanium alloy. Mater Des 31(3):1650–1655
45. Zhou L, Liu H J, Liu QW (2010) Effect of rotation speed on microstructure and mechanical properties of Ti–6Al–4V friction stir welded joints. Mater Des (1980–2015) 31(5):2631–2636
46. Matsushita M, Kitani Y, Ikeda R et al (2012) Microstructure and toughness of friction stir weld of thick structural steel. ISIJ Int 52(7):1335–1341
47. Fujii H, Cui L, Tsuji N et al (2006) Friction stir welding of carbon steels. Mater Sci Eng, A 429(1/2):50–57
48. Ozekcin, Jin, Koo et al (2004) A microstructural study of friction stir welded joints of carbon steels. Int J Offshore Polar Eng 14(4):284–288
49. Cui L, Fujii H, Tsuji N et al (2007) Friction stir welding of a high carbon steel. Scripta Mater 56(7):637–640
50. Chung YD, Fujii H, Ueji R et al (2010) Friction stir welding of high carbon steel with excellent toughness and ductility. Scripta Mater 63(2):223–226
51. Ghosh M, Kumar K, Mishra RS (2010) Analysis of microstructural evolution during friction stir welding of ultrahigh-strength steel. Scripta Mater 63(8):851–854
52. Xue P, Xiao BL, Wang WG et al (2013) Achieving ultrafine dual-phase structure with superior mechanical property in friction stir processed plain low carbon steel. Mater Sci Eng, A 575:30–34
53. Xue P, Ma ZY, Komizo Y et al (2016) Achieving ultrafine-grained ferrite structure in friction stir processed weld metal. Mater Lett 162:161–164
54. Xue P, Zhang XX, Wu LH et al (2016) Research progress on friction stir welding and processing. Acta Metall Sin 52(10):1222–1238 (in Chinese)
55. Nandan R, Roy GG, Lienert TJ et al (2006) Numerical modelling of 3D plastic flow and heat transfer during friction stir welding of stainless steel. Sci Technol Weld Joining 11(5):526–537
56. Sato YS, Nelson TW, Sterling CJ (2005) Recrystallization in type 304L stainless steel during friction stirring. Acta Mater 53(3):637–645
57. Du DX, Fu RD, Li YJ et al (2014) Gradient characteristics and strength matching in friction stir welded joints of Fe–18Cr–16Mn–2Mo–0.85N austenitic stainless steel. Mater Sci Eng: A 616:246–251
58. Han J, Li HJ, Zhu ZX et al (2014) Microstructure and mechanical properties of friction stir welded 18Cr–2Mo ferritic stainless steel thick plate. Mater Des 63:238–246
59. Okamoto H (1882) Desk handbook: phase diagrams for binary alloys. ASM International, Almere
60. Kostka A, Coelho RS, dos Santos J et al (2009) Microstructure of friction stir welding of aluminium alloy to magnesium alloy. Scripta Mater 60(11):953–956
61. Xue P, Ni DR, Wang D et al (2011) Effect of friction stir welding parameters on the microstructure and mechanical properties of the dissimilar Al–Cu joints. Mater Sci Eng, A 528(13/14):4683–4689
62. Xue P, Xiao BL, Ma ZY (2015) Effect of interfacial microstructure evolution on mechanical properties and fracture behavior of friction stir-welded Al-Cu joints. Metall and Mater Trans A 46(7):3091–3103
63. Wei YN, Li JL, Xiong JT et al (2012) Microstructures and mechanical properties of magnesium alloy and stainless steel weld-joint made by friction stir lap welding. Mater Des 33:111–114
64. Zhang ZK, Wang XJ, Wang PC et al (2014) Friction stir keyholeless spot welding of AZ31 Mg alloy-mild steel. Trans Nonferrous Metals Soc China 24(6):1709–1716
65. Wang D, Xiao BL, Ma ZY et al (2009) Friction stir welding of $Zr_{55}Cu_{30}Al_{10}Ni_5$ bulk metallic glass to Al–Zn–Mg–Cu alloy. Scripta Mater 60(2):112–115
66. Sun YF, Fujii H (2013) Microstructure and mechanical properties of dissimilar spot friction stir welded $Zr_{55}Cu_{30}Al_{10}Ni_5$ bulk metallic glass to pure copper. Intermetallics 33:113–119

67. Liu FC, Liao J, Nakata K (2014) Joining of metal to plastic using friction lap welding. Mater Des (1980–2015) 54:236–244
68. Goushegir SM, dos Santos JF, Amancio-Filho ST (2014) Friction Spot Joining of aluminum AA2024/carbon-fiber reinforced poly(phenylene sulfide) composite single lap joints: microstructure and mechanical performance. Mater Des (1980–2015) 54:196–206
69. Meisnar M, Bennett JM, Andrews D et al (2019) Microstructure characterisation of a friction stir welded hemi-cylinder structure using Ti-6Al-4V castings. Mater Charact 147:286–294
70. Kim H, Lee K, Kim J et al (2020) A study on the friction stir welding experiment and simulation of the fillet joint of extruded aluminum material of electric vehicle frame. Appl Sci 10(24):9103
71. Ahmed MMZ, El-Sayed Seleman MM, Touileb K et al (2022) Microstructure, crystallographic texture, and mechanical properties of friction stir welded mild steel for shipbuilding applications. Materials 15(8):2905
72. Shen ZK, Yang XQ, Zhang ZH et al (2013) Microstructure and failure mechanisms of refill friction stir spot welded 7075–T6 aluminum alloy joints. Mater Des 44:476–486
73. Labus Zlatanovic D, Balos S, Bergmann JP et al (2020) An experimental study on lap joining of multiple sheets of aluminium alloy (AA 5754) using friction stir spot welding. Int J Adv Manuf Technol 107(7):3093–3107
74. Karthikeyan R (2022) Establishing relationship between optimised friction stir spot welding process parameters and strength of aluminium alloys. Adv Mater Process Technol 8(1):1173–1195
75. Ojo OO, Taban E, Kaluc E (2015) Friction stir spot welding of aluminum alloys: a recent review. Mater Test 57(7/8):609–627
76. Shah PH, Badheka VJ (2019) Friction stir welding of aluminium alloys: an overview of experimental findings–process, variables, development and applications. Proc Inst Mech Eng, Part L: J Mater: Des Appl 233(6):1191–1226
77. Buffa G, Fratini L, La Commare U et al (2022) Joining by forming technologies: current solutions and future trends. Int J Mater Form 15(3):27
78. Prabhakar DAP, Shettigar AK, Herbert MA et al (2022) A comprehensive review of friction stir techniques in structural materials and alloys: challenges and trends. J Market Res 20:3025–3060
79. Geng PH, Ma YW, Ma NS et al (2022) Effects of rotation tool-induced heat and material flow behaviour on friction stir lapped Al/steel joint formation and resultant microstructure. Int J Mach Tools Manuf 174:103858
80. Chitturi V, Pedapati SR, Awang M (2019) Effect of tilt angle and pin depth on dissimilar friction stir lap welded joints of aluminum and steel alloys. Materials 12(23):3901
81. Palanivel S, Nelaturu P, Glass B et al (2015) Friction stir additive manufacturing for high structural performance through microstructural control in an Mg based WE43 alloy. Mater Des (1980–2015) 65:934–952

Chapter 5
Narrow Gap Welding

5.1 Principle, Characteristics and Classification of Narrow Gap Welding

With the complexity of working conditions, the application of large and thick plates in engineering structures is more and more extensive, and the welding problem of large and thick plates has gradually attracted the attention of researchers. At present, the large and thick plates welding encountered in the manufacturing of thick wall structures are mainly submerged arc welding, electroslag welding and multi-layer multi-pass welding processes. However, the first two methods inevitably adopt the welding process parameters of large heat input during welding, which often leads to serious embrittlement of the welded joints, especially the heat impact zone, which is not allowed in some occasions with high impact toughness requirements. At the same time, when submerged arc welding and electroslag welding are used, U-shaped or X-shaped grooves are inevitably used, which not only wastes raw materials, but also increases the processing difficulty of raw materials. Using multi-layer and multi-pass welding, not only is the welding efficiency low, but also the welding quality is often difficult to guarantee. Therefore, it is urgent to research and develop efficient and high-quality new welding technology for large and thick plates.

The aim of many process developments has been to decrease the time taken to complete the joint (so reducing labour costs) by increasing metal deposition rates and using automation. An alternative approach is to reduce the weld size or joint volume. In fillet welds the possibility of a reduction in weld size will depend on design constraints and the achievement of smaller weld volume is easily controlled by the operating parameters of the process. In butt welds significant changes in weld metal volume may require modification of the joint configuration, a change of process, or both.

R. Li and T. Li, *Advanced Welding Methods and Equipment*,
https://doi.org/10.1007/978-981-97-4109-0_5

5.1.1 Definition and Characteristics of Narrow Gap Welding

Since the first publication of the narrow gap welding technology developed by the Battelle Institute in the Iron Age magazine in December 1963, narrow gap welding technology immediately received a lot of attention from welding experts around the world as a more advanced welding technology.

In conventional "V" preparations, the joint volume and, hence, weld completion time, increases dramatically in proportion to the square of thickness (Fig. 5.1). As the angle of the preparation is reduced, the weld metal volume and joint completion rate decrease and, if a narrow parallel-sided gap is used, the difference becomes significant, particularly on thicker sections. These narrow parallel sided gaps or square closed-butt preparations are inherent in certain processes, such as flash butt, friction, MIAB, plasma keyhole, laser and electron beam welding. Special techniques must be used, however, to allow the use of narrower gaps with the conventional arc welding processes. In addition to the potential improvement in welding economics which narrow-gap techniques provide, they can also give reduced distortion and more uniform joint properties.

Malin [1] has compiled the following characteristics of narrow gap welding from many publications: (1) It is a special technique that utilizes existing arc welding methods; (2) I-shaped grooves are mostly used, with the size of the bevel angle depending on the amount of deformation in the weld; (3) multi-layer welding; (4) the same number of passes (usually 1 or 2) for each layer from the bottom up; (5) welding with small or medium heat input; and (6) the possibility of all-position welding.

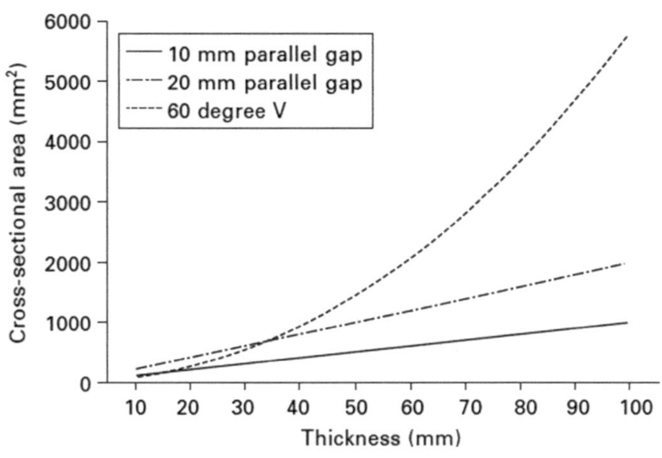

Fig. 5.1 Effect of joint preparation on weld cross-sectional area

5.1.2 Advantages and Disadvantages of Narrow Gap Welding

Figure 5.2 shows a comparative diagram of the welded joints with a plate thickness of 200 mm under three welding methods. It can be seen that narrow gap welding is a multi-layer single-pass welding process. The cross-section of the narrow gap welding is the best with small heat-affected zone. Submerged arc welding is multi-layer multi-pass welding process with a large heat-affected zone. The efficiency of submerged arc welding is the lowest and the performance of the welded joint is not as good as the welded joint of narrow gap welding. Electroslag welding is a single layer single pass welding process. The efficiency of electroslag welding is definitely the highest. However, the heat input of electroslag welding is too large, resulting in a large heat-affected zone and the worst joint performance.

From the above comparison, it is clear that narrow gap welding is a welding method with high welding quality, minimal distrotion, and high operational efficiency. The following advantages are available in comparison with commonly used welding processes.

(1) Reduced bevel angles mean that thick sections can be welded more economically. A standard pipe bevel angle is 37.5°, whereas a narrow gap weld angle can be as little as 3°.
(2) Weld shrinkage is controlled due to reduced weld layers.
(3) Joint volume and requested consumbles are reduced.
(4) Increased rate of joint completion means reduced welding time and labour costs.
(5) Distortion of angle is reduced, due to almost parallel joint side walls.
(6) The low heat input reduces stress on the joined metal.
(7) Metallurgical properties are more consistent.

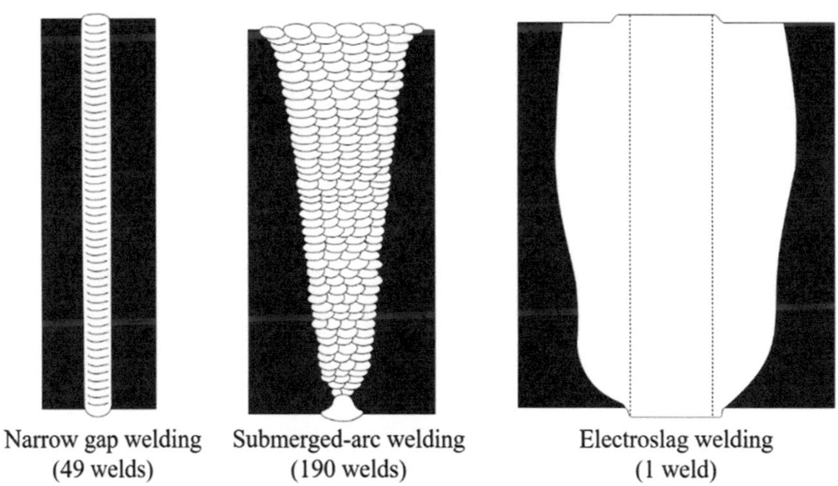

Fig. 5.2 Comparison of joint shapes of several typical welding methods

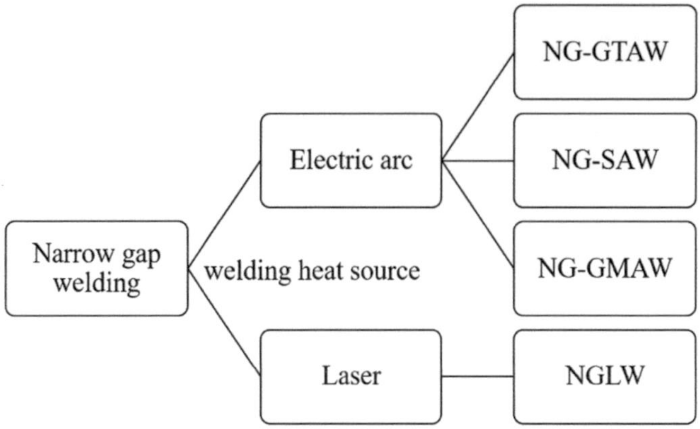

Fig. 5.3 Classification of narrow gap welding

However, narrow gap welding does require specialised equipment, because of the limited accessibility to the root of the preparation. The disadvantages of the narrow gap technique are:

(1) The weld is more prone to defects for certain welding processes—especially lack of sidewall fusion.
(2) It is difficult to remove any defects when detected, because of poor joint accessibility.

5.1.3 Classification of Narrow Gap Welding

Narrow gap welding technology is classified according to the process adopted (Fig. 5.3) [2], which can be divided into narrow gap tungsten arc welding (NG-GTAW), narrow gap submerged arc welding (NG-SAW), narrow gap gas metal arc welding (NG-GMAW), narrow gap laser welding (NGLW), and so on. Each has its own characteristics and scope of adaptation.

5.2 Narrow Gap GTAW

Gas Tungsten Arc Welding (GTAW) technology is a very versatile metal joining process with stable welding process and low heat input. The GTAW process is easily adapted for narrow-gap operation since compact welding torches are available and, in thinner sections, it is only necessary to extend the distance by which the electrode projects from a conventional gas shroud (Fig. 5.4). Some provision must also be made for feeding a filler wire into the weld pool, but, again, conventional equipment is often suitable for thicknesses up to 12 mm. For thicker sections, a special torch design is required to ensure adequate gas cover and reliable wire positioning. Torches specifically designed for narrow-gap GTAW are usually elongated as shown in Fig. 5.5, the shielding gas is delivered to rectangular slots either side of the electrode and, in addition, gas may be supplied through holes in the side of the blade. Whilst the design is, in principle, fairly straightforward, it has been found that the achievement of adequate gas cover without entrainment of air from the surroundings is quite difficult. Telescopic gas shrouds and flexible surface baffles are often added to supplement the gas shroud. The process has been used for welding thicknesses of up to 200 mm, particularly for circumferential welds and pipe-work.

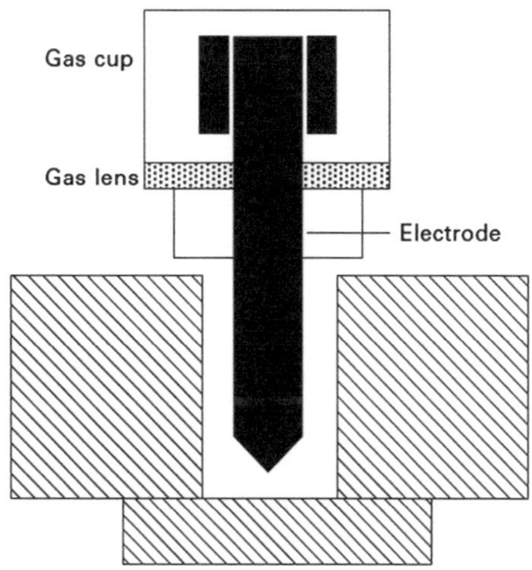

Fig. 5.4 Narrow-gap welding with a conventional GTAW torch [3]

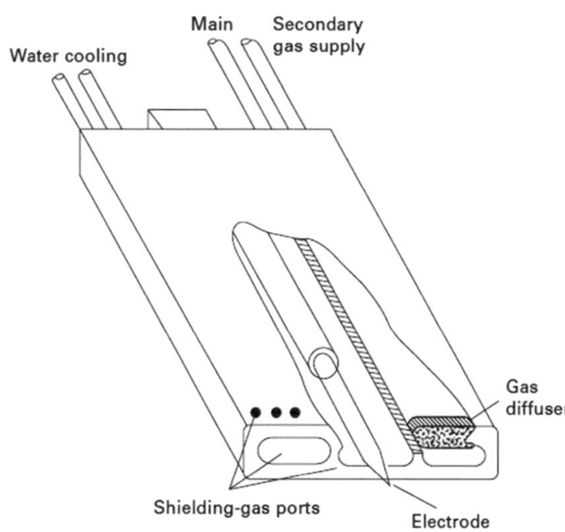

Fig. 5.5 Narrow-gap GTAW torch [3]

The hot wire GTAW process is particularly suitable for narrow-gap welding, since it allows the deposition rate of the process to be increased to levels similar to those of GMAW. Many commercial applications of narrow-gap GTAW employ hot wire addition and an interesting example is the Roboweld™ system which was designed for welding transmission pipelines both on land and on pipe lay barges. Seam tracking and accurate positioning of the electrode may be achieved, but some systems also include facilities for video monitoring of the arc and torch position with the ability for the operator to override the automatic control to correct the torch position.

Due to its small molten pool, the lack of sidewalls fusion was obvious and impeded the development of narrow gap GTAW welding. Cook and Levick [2] proposed the hot wire GTAW welding process to solve the lack of sidewall fusion defects, avoid the formation of welding cracking and improve weld efficiency. However, the occurrence of magnetic arc blow due to wire current made the welding process become unstable. Pulsed current was proposed to preheat the filler wire to avoid the magnetic blow. The magnetic blow produced during only the wire pulsed current period and no magnetic blow occurred during non-pulse period. The ratio of the magnetic blow period became reduced and the workability was not so much damaged as the ratio of this pulse period decreased.

5.3 Narrow Gap GMAW

Narrow gap gas metal arc welding (NG-GMAW) was successfully developed after 1975. This process is realized after resolving the fusion issue on the groove sidewalls by utilizing a specially designed wire bending structure to maintain wire curvature. NG-GMAW employs arc oscillation to reach both sides of the welded steel plates. This method utilizes arc rotation to expand the arc's heating area, increase the arc's heating time on the sidewalls, and thereby ensure thorough sidewall fusion. Figure 5.6 shows the schematic diagram of oscillation systems for NG-GMAW. These systems are relatively complex and the minimum gap width is often limited by the need to move the whole welding head. Alternative systems that rely on controlled deformation of the filler wire as it passes through the feeding system have therefore been developed. These devices use bending rollers to introduce a wave-like or spiral deformation into the wire in order to cause controlled oscillation of the arc. Although the mechanism required to produce the deformation may be complex, the part of the torch that enters the joint is compact and satisfactory joints can be made in gap widths down to around 9 mm.

In pursuit of enhanced sidewall fusion and improved weld formation, numerous specific forms of NG-GMAW applications have been developed. These classifications are primarily shown in Fig. 5.7. NG-GMAW can be categorized into single-wire welding and multi-wire welding based on the number of wires used, with dual-wire welding being most commonly employed in multi-wire scenarios. In single-wire applications, arc forms include oscillating, rotating, and non-oscillating arcs. In

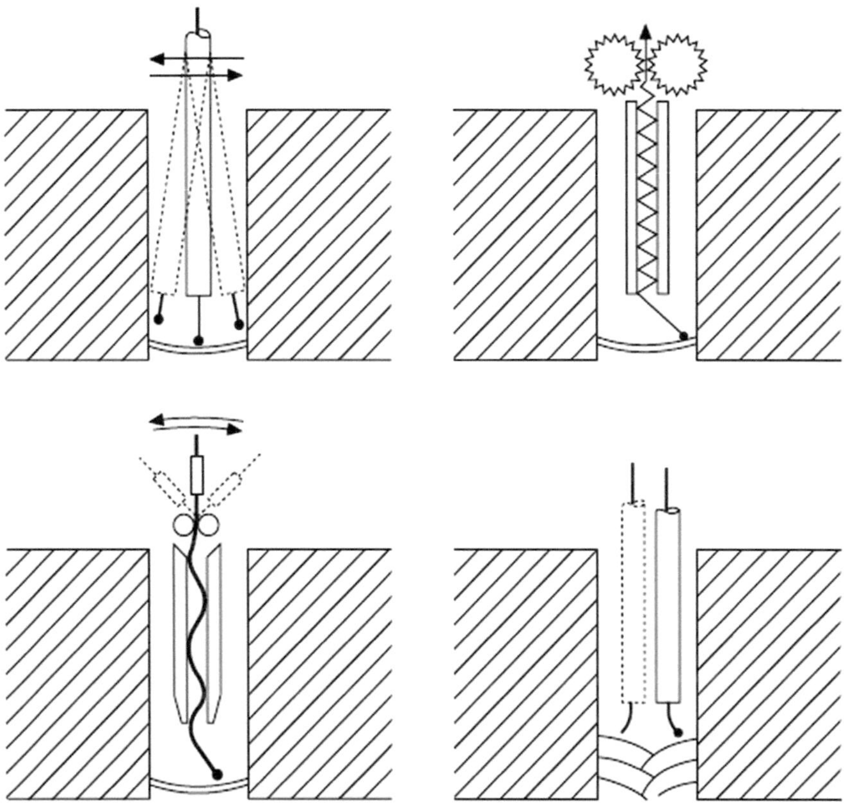

Fig. 5.6 Oscillation techniques for narrow-gap GMAW

recent years, the development of ultra-narrow gap welding with gaps below 5 mm has emerged.

The dual-wire NG-GMAW method was initially proposed by the United States Battle Research Institute to address the welding challenges of submarine hull circumferential seams. As the single-wire arc has a limited heating range that cannot ensure adequate sidewall fusion, dual-wire welding was adopted. This approach involves bending the welding wires or manipulating the contact tip to direct the two wires towards the root of the gap on both sides.

Another method to expand the heating range of a single-wire arc is arc oscillation (swinging), similar to the slight pauses during manual welding when moving the arc to both sides of the groove. This not only widens the heating area but also increases the arc's heating time on both sidewalls, thus addressing poor sidewall fusion in narrow gap GMAW. Methods for inducing arc oscillation include torch swinging, wire bending, utilizing alternating magnetic fields, and manipulating the contact tip.

NG-GMAW boasts advantages such as high deposition efficiency, low cost, no slag removal requirement, all-position welding capability, and ease of automation.

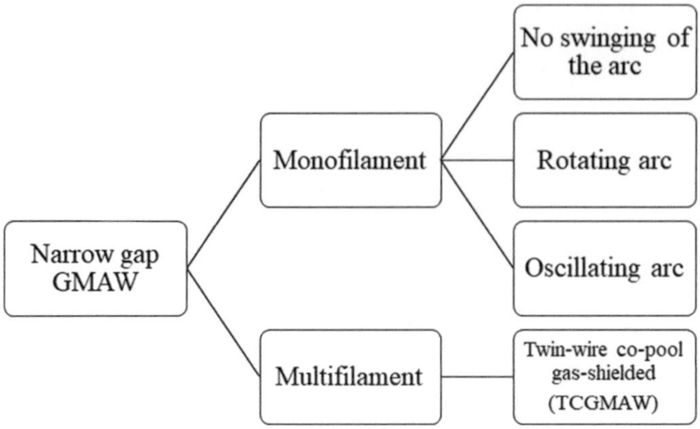

Fig. 5.7 Classification of narrow gap GMAW

Focused on ensuring sidewall fusion, various new methods, techniques, and equipment have emerged. With ongoing in-depth research by domestic and international scholars on NG-GMAW, as well as the further popularization of GMAW methods in our country, it is believed that narrow gap GMAW will have broader prospects for application.

5.4 Narrow Gap SAW

Narrow Gap Submerged Arc Welding (NG-SAW) emerged in the 1980s and was quickly adopted for industrial production. Its primary application areas include thick-walled low-alloy steel vessels and other heavy welding structures. Figure 5.8 shows the schematic diagram of NG-SAW. Welded joints produced using NG-SAW exhibit higher resistance to delayed cracking. The strength and impact toughness of NG-SAW joints surpass those of traditional wide bevel submerged arc welding joints. Compared to conventional submerged arc welding, NG-SAW can increase overall efficiency by 50–80%. It also allows for a reduction of 38–50% in wire consumption and 56–64.7% in flux usage. Comprehensive sets of equipment for single-wire, dual-wire, and multi-wire NG-SAW have been developed, primarily for horizontal or near-horizontal welding positions. These applications require flux to carry current and slag removal, ensuring adequate mechanical performance of the welds.

NG-SAW technology fully inherits the inherent advantages of the SAW method while to some extent overcoming certain limitations of traditional SAW methods. The primary technical advantages of NG-SAW are as follows:

(1) High deposition rate.
(2) Low sensitivity of weld geometry dimensions to fluctuations in welding current and voltage.

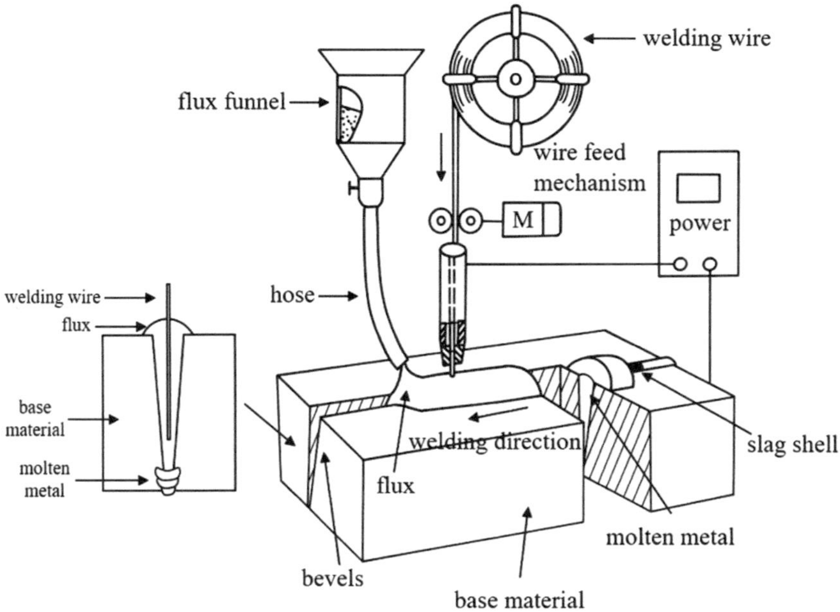

Fig. 5.8 Schematic diagram of narrow gap submerged arc welding

In other arc welding methods, fluctuations in welding current and voltage beyond 10% often lead to noticeable changes in weld geometry parameters, and even result in process defects like lack of fusion, incomplete penetration, and undercutting. Under NG-SAW conditions, even with fluctuations of ± 10% in welding current and voltage, the variation in weld geometry parameters is not significantly pronounced. The likelihood of defects such as lack of fusion and incomplete penetration is low, indicating that weld quality is minimally affected by fluctuations in welding current and voltage. This technical advantage of traditional SAW methods is completely inherited by NG-SAW, primarily due to three factors: high current intensity used in SAW, typically 2–3 times that of other arc welding methods; higher thermal efficiency of the arc under the flux layer; and greater allowance in bead width and depth compared to other arc welding methods, resulting in stable and reliable welding process quality.

(3) Enhanced weld appearance quality.

A significantly higher weld shape factor (W/H), slag wall transition, longer liquid presence time in the weld pool, and comprehensive protection from the molten flux all contribute to a smoother and more aesthetically pleasing appearance of NG-SAW welds.

The main technical limitations of NG-SAW are as below:

(1) Unsuitability for welding positions other than flat welding, especially vertical-up, vertical-down, and all-position welding.

(2) The reduction in the fill area of welding bevel under the same plate thickness is smaller due to slag removal constraints. This reduction is at least 30% larger than the fill area of solid wire narrow gap GMAW. This limitation becomes more prominent when welding thick and ultra-thick plates.
(3) Welding heat input is significantly higher than that of narrow gap GMAW. The weld joint's plasticity, toughness, and strength parameters are not as favorable as those achieved with lower heat input narrow gap GMAW. This limitation is particularly pronounced with large-diameter wires and multi-wire narrow gap submerged arc welding. Welded joints made of low-alloy high-strength steel are often not suitable for direct service load and require post-weld heat treatment.

Multi-wire narrow gap submerged arc welding is an advanced welding technique widely employed in heavy industries. This section delves into the aspects of welding torch design, characteristics and droplet transition, welding formation, and applications of multi-wire narrow gap SAW welding. Figure 5.9 shows a schematic diagram of narrow gap thriple wire submerged arc welding. It is extremely important that the arrangement of both wires travelling parallel with both narrow-gap fusion faces is optimum. This includes an optimum distance between the wires, an optimum wire diameter and an optimum angle between the wires when they are leaving the contact tube. The wire travelling as the first in the welding direction serves only to fill the centre of the weld.

The welding torch design plays a crucial role in ensuring the stability and efficiency of multi-wire narrow gap SAW welding. Several factors need consideration in the design process:

(1) Wire Arrangement: The arrangement of multiple wires directly impacts welding formation and seam quality. Typically, a parallel arrangement of wires is utilized, with optimized wire spacing based on welding material, thickness, and requirements to ensure proper fusion of the wires and the base metal.
(2) Wire Feeding Mechanism: As multi-wire welding demands individual control of wire feeding speed and current, the wire feeding mechanism should possess independent control functionality to cater to the intricate welding parameter demands.
(3) Flux Coating Spray System: An adequate flux coating spray system protects the welding area, reduces oxidation, improves pool stability, and lowers the occurrence of porosity and slag entrapment.

Concerning droplet transition, the multi-wire involvement during welding makes droplet transition more complex. Proper control of current, welding speed, wire spacing, and flux spraying are critical to ensuring a stable droplet transition process.

Multi-wire narrow gap SAW welding exhibits the following characteristics:

(1) High Efficiency: The welding of multiple wires enables a larger weld bead width and faster welding speed, thereby enhancing overall production efficiency.
(2) High-Quality Weld Seam: Multi-wire welding provides a larger welding pool, facilitating the complete fusion of wires and base metal, leading to high-quality weld seams.

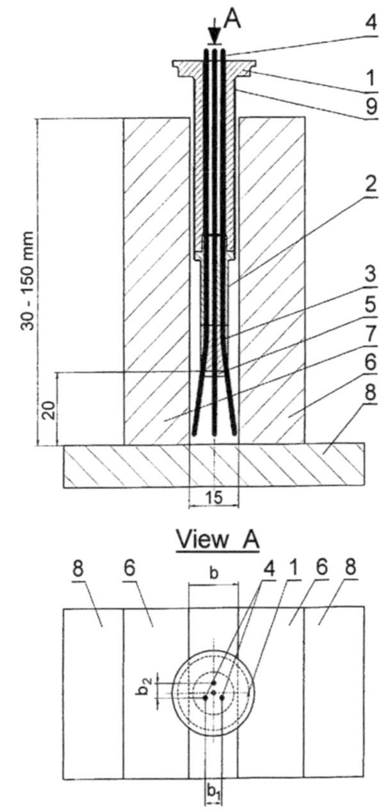

Fig. 5.9 Narrow-gap submerged arc welding with a triple wire electrode

(3) Low Heat Input: Due to the increased weld bead width, the heat input per unit length is relatively lower, reducing the risk of welding distortion and stress concentration.

Multi-wire narrow gap SAW welding, through the arrangement of multiple and parallel wires, enables the realization of wider weld beads, allowing for the completion of relatively broad weld seams in a single welding pass. This formation method is particularly suitable for thick plate welding and transverse welding, significantly improving welding efficiency and quality.

Multi-wire narrow gap SAW welding finds extensive applications in the following areas:

(1) Shipbuilding: Welding of ship structures and plates with long weld seams benefits from the increased welding efficiency and reduced costs provided by multi-wire narrow gap SAW welding.
(2) Pressure Vessels: Multi-wire narrow gap welding is well-suited for manufacturing thick-walled pressure vessels, ensuring weld seam quality and vessel safety.

(3) Bridge Construction: Multi-wire narrow gap SAW welding enhances welding speed and is suitable for welding bridge components.
(4) High-Strength Steel Structures: For high-strength steel welding, multi-wire narrow gap SAW welding reduces welding distortion and the size of the heat-affected zone.

Multi-wire narrow gap SAW welding is an efficient and high-quality welding method. Through optimized welding torch design, controlled characteristics and droplet transition, and favorable welding formation advantages, it has found widespread use in shipbuilding, pressure vessel manufacturing, bridge construction, and high-strength steel structures. With continuous technological advancements, the significance of multi-wire narrow gap SAW welding in industrial manufacturing will continue to grow, contributing to the development of various heavy industries.

5.5 Narrow Gap Laser Welding

Laser welding is a very flexible process wherein the laser beam can be transmitted through optical fibres to inaccessible locations for fabrication of thick-section steels. Currently, laser welding refers to single-pass autogenous operations. Autogenous laser welding is performed by localised melting of the mating surfaces using laser beam and allowing solidification of the base metal without the addition of any other material. In comparison to arc welding, laser welding is economical, reliable, and a versatile joining method for most materials including shape memory alloys. Moreover, the thermal energy imparted into the weld zone can be precisely controlled by monitoring the laser beam properties and the welding traverse speeds, thereby having the capability to tune the characteristics of the weld joint with small heat-affected zone (HAZ) [4]. Furthermore, laser-welded joints tend to retain functional performance of the base materials, rendering the welded structure to perform reliably during service [5].

Autogenous laser welding can be typically employed to weld plates of thicknesses in the order of 1–2 mm/kW of the available laser power [6]. Several research studies including Katayama et al. [7] have proven successful single-pass laser welding of 70-mm-thick 304 stainless steel, however, using very high laser powers, very low welding speeds, and low atmospheric pressures, which is unrealistic for industrial production. If the heat input from the laser beam during welding is not optimised, it can result in the formation of unwanted defects. To optimise the process, conduction mode welding is preferable where the aspect ratio of the fusion zone is ≈ 1, compared to keyhole mode welding where the aspect ratio is $\gg 1$. Moreover, laser welding process has several other drawbacks including stringent requirements for setup accuracy and very expensive welding systems that can incorporate lasers of high powers. Therefore, narrow gap laser welding (NGLW) with added filler material, utilisation of available laser power, and the use of multi-pass non-autogenous welding has tremendous potential for application when fabricating thick steel plates, see Fig. 5.10. In NGLW, a laser beam is directed into a narrow groove of the heavy section along with filler wire through a guide for layer-by-layer welding. The addition of filler material not only enhances the quality of the weld but can also be effectively used to tailor the microstructural and mechanical properties of the fabricated parts. The interaction between the laser beam and filler wire is critical to the stability of the NGLW process [8]. A trailing shield gas is also administered during the process to avoid any unwanted atmospheric contamination.

Thick-section steels are the most commonly used materials in heavy industries including power and nuclear plants, shipbuilding and submarine manufacturing, and offshore pipelines. Moreover, structural steels are preferred for these applications due to their metallurgical stability, high corrosion resistance, and good creep and ductility at elevated temperatures. For example, most of the plumbing inside nuclear reactors and intermediate-level nuclear waste storage sites is made of stainless steels. Typically, arc welding processes are employed for joining thick-section steels. However, these processes have low production efficiencies as they generally involve low welding speeds and high number of weld passes.

Laser welding and electron beam welding techniques have received more attention for fabricating thick-section steels owing to their high welding efficiencies and low distortion for very narrow deep penetration welds. Although there are cases where

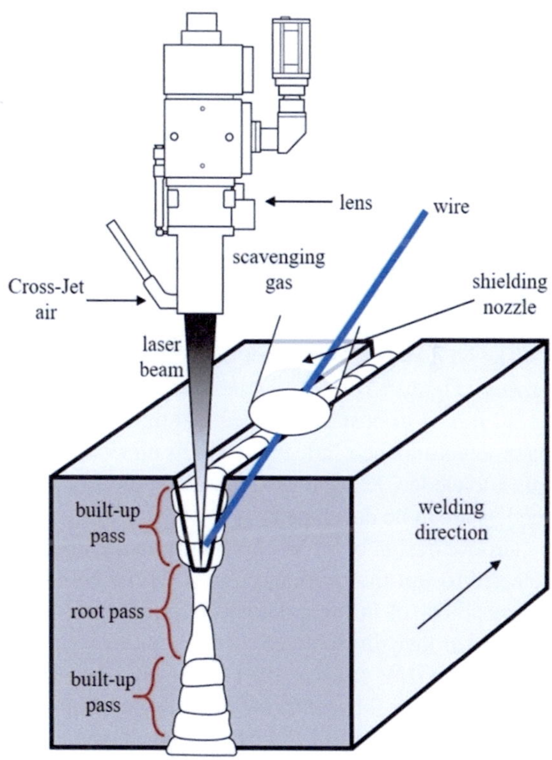

Fig. 5.10 Schematic diagram of narrow gap laser welding

electron beam welding has been proven to weld plates of thicknesses more than 50 mm and sometimes exceeding 100 mm in a single weld pass [9, 10], the process requires vacuum or low-pressure inert conditions with stringent gap tolerances which is not ideal for several fabrication applications including shipbuilding and pipeline industries [4].

NGLW technique has several advantages over conventional multi-pass arc welding processes. One key advantage is that it allows higher welding speeds owing to its high power density, thereby resulting in increased productivity. Moreover, the NGLW process is economical compared to other welding processes when fabricating thick-section steel structures. Consequently, the distortion and residual stresses are lower due to low heat input [8]. Shi et al. [11] reported an angle distortion of only 0.045 rad for a 20-mm-thick austenitic stainless steel plate after NGLW. Although NGLW is a multi-pass welding method, the number of passes required to finish a particular job is considerably reduced in comparison to arc welding of the same job. Moreover, the technique produces a narrow heat affected zone due to the very narrow bead width [12]. Guo et al. [13] found that the cumulative heat input and the amount of filler wire consumed for a GMAW joint was ~ 6 times that of the NGLW joints; however, the welding speed of the NGLW process was ~ 1.5 times that of the GMAW process.

In brief, Ning et al. [14] have shown that with an optimized set of NGLW process parameters, it is possible to achieve welding efficiency 3.75 times that of the industrially employed gas tungsten arc welding (GTAW). Other than steels, NGLW has also been applied to other materials such as aluminum alloys [13, 15], nickel superalloys [16–18], and titanium-steel bimetallic sheets [19].

NGLW typically is a multi-pass multi-layer process which involves four basic operation steps. Firstly, a groove design with parallel or near-parallel side-walls appropriate to the thickness of the plates is identified and machined. This is followed by autogenous root pass welding to create a backing weld. The remainder depth of the groove is welded with laser welding and the addition of filler material as illustrated in Fig. 5.11. As it is a relatively easy to implement process, there is a high potential for automation with increased welding productivity, provided the process is well-developed.

The NGLW process is a bit more complicated than the autogenous laser welding process. Figure 5.12 demonstrates the process parameters involved in an NGLW operation. It is of utmost importance that the correlations between the different process parameters and their influential effects on weld quality are well understood. To attain this knowledge, extensive experimentation is required as well as thermomechanical models are to be developed.

Groove design of the workpieces is an important parameter that should be ideally selected to suit the welding process. In the NGLW process, one must consider the practicability of fitting a defocused laser beam as well as the filler wire within the selected groove dimensions.

In the NGLW process, the type of filler material has a significant impact on the quality of the weld seams. Ideally, the chemical compositions of the filler wire are near-identical to that of the base metal. In some instances, it is required to change the microstructure of the weld zone using filler wire of different compositions [20].

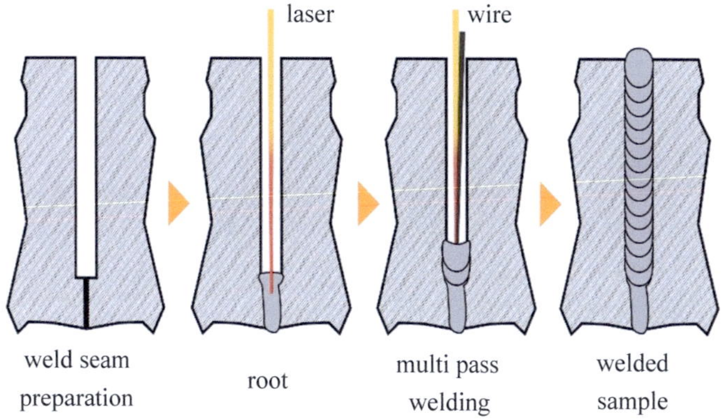

Fig. 5.11 Sequence of process steps for a multi-pass NGLW

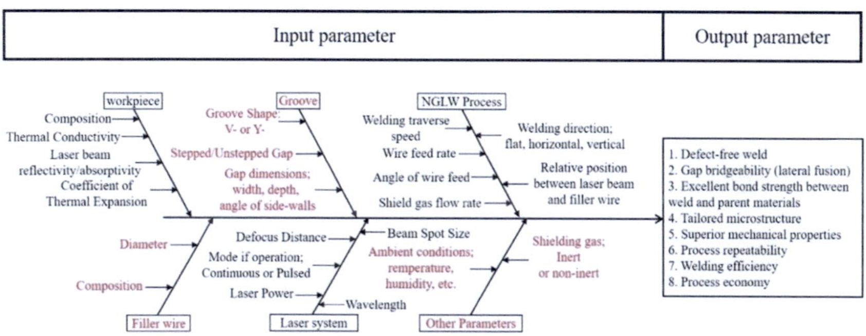

Fig. 5.12 Fishbone diagram representing input and output parameters involved in an NGLW process

The industrial demand for improving the quality and productivity of the welding processes has resulted in the development of advanced welding technologies as well as welding strategies.

Although arc welding processes such as gas-metal arc welding [21], gas-tungsten arc welding [22] and submerged arc welding [23] are industrially accepted fabrication techniques for up to 50 mm-thick steel structures, they are riddled with a number of technological limitations such as large weld and HAZ regions with high residual thermal strains, susceptibility to hot cracking which necessitates post-weld tempering heat treatment, and difficulty in maintaining stable arcing over the entire weld seam [24]. This leads to reduced productivity due to the frequent stop and start welding sequences. Narrow gap laser welding is one such unique technique that can be used to weld steels up to 100 mm thick with relatively low heat input, increased productivity owing to high welding speeds utilized, reduced consumption of consumables including filler material and inert gases used for shielding, thereby resulting in greater cost savings, and also offering better control of weld properties.

However, NGLW of thick-section structural steels has not been widely adopted for practical applications due to limited understanding of the process as well as several prevalent research gaps such as process control and monitoring, technological adaptability, applicability to different structural designs, elimination of process-induced defects, and robust assessment of structural integrity. Moreover, a major limitation of the NGLW process that also needs to be critically investigated is the occurrence of defects such as lack of fusion, pores, and slag inclusions.

5.6 Summary and Prospects

The use of narrower joint gaps and reduced preparation angles can result in significant improvements in productivity. The use of processes which involve the inherent use of a narrow gap (EBW, laser, plasma, friction, MIAB) automatically exploits these advantages, whilst systems have been developed to allow narrow gaps to be used with GTAW, GMAW and SAW processes. Optimization of welding parameters and in-process control are essential to avoid defects in narrowgap applications; the restricted access of the gap will make progressive repair difficult, but good procedure control should obviate these problems and the use of narrow-gap welding may be seen as a way of imposing a reasonable level of discipline into the control of welding operations.

With the evolution of automation and intelligence technologies, narrow gap welding will increasingly rely upon cutting-edge robotics, sensors, and real-time monitoring systems to achieve precision control and adaptive adjustments, and improve production efficiency and welding quality. With the wide application of multi-material structures in manufacturing, narrow gap welding will realize high-quality welding between different materials and promote innovative combinations of dissimilar materials. Besides, it might also be promising to develop in the fields of micro-size welding and environmental protection and sustainability.

Problems

1. What is narrow gap welding? Provide a brief description of its differences from conventional welding methods.
2. Explain the main principles of narrow gap welding and its role in the welding process.
3. Based on different gap sizes, how can narrow gap welding be classified? Please list and briefly describe them.
4. What are the advantages and limitations of narrow gap welding compared to traditional welding methods?
5. Explain the basic working principle of narrow gap arc welding, along with its applicable materials and fields of use.
6. What are the advantages of narrow gap laser welding compared to narrow gap arc welding? List at least two points.

7. What is the core concept of narrow gap laser-arc hybrid welding? Briefly describe its working principle.
8. What are the main characteristics of ultra-narrow gap welding? Why is ultra-narrow gap welding chosen in certain situations?
9. In what situations do you think narrow gap arc welding is more suitable than conventional arc welding? Please provide reasons for your answer.
10. Briefly outline the working steps of narrow gap laser welding and explain its pros and cons.
11. How does the narrow gap laser-arc hybrid welding achieve complementary advantages through the combination of these two welding methods? Provide an example.
12. What is the significance of applying ultra-narrow gap welding in the aerospace field? Provide a concise explanation.
13. Why is narrow gap welding the preferred method for some demanding welding tasks, such as thick plate connections? Explain your reason.
14. Does narrow gap welding demand higher operational skills and equipment requirements? Please show your viewpoint.

References

1. Malin VY (1983) The state-of-the-art of narrow gap welding. Weld J 62(4):22–32
2. Tsuchiya K, Kawamura H, Kalinin G (2000) Re-weldability tests of irradiated austenitic stainless steel by a TIG welding method. J Nucl Mater 283–287(Part II):1210–1214
3. Norrish J (2006) Advanced welding processes. Woodhead Publishing, London
4. Yu YC, Yang SL, Yin Y et al (2013) Multi-pass laser welding of thick plate with filler wire by using a narrow gap joint configuration. J Mech Sci Technol 27(7):2125–2131
5. Zeng Z, Yang M, Oliveira JP et al (2016) Laser welding of NiTi shape memory alloy wires and tubes for multi-functional design applications. Smart Mater Struct 25(8):085001
6. Oliveira JP, Schell N, Zhou N et al (2019) Laser welding of precipitation strengthened Ni-rich NiTiHf high temperature shape memory alloys: microstructure and mechanical properties. Mater Des 162:229–234
7. Oliveira JP, Zeng Z, Berveiller S et al (2018) Laser welding of Cu-Al-Be shape memory alloys: microstructure and mechanical properties. Mater Des 148:145–152
8. Ramakrishna RVSM, Amrutha PHSL, Rahman Rashid RA et al (2020) Narrow gap laser welding (NGLW) of structural steels—a technological review and future research recommendations. Int J Adv Manuf Technol 111(7):2277–2300
9. Zhang XD, Ashida E, Tarasawa S et al (2011) Welding of thick stainless steel plates up to 50 mm with high brightness lasers. J Laser Appl 23(2):022002
10. Feng JC, Rathod DW, Roy MJ et al (2017) An evaluation of multipass narrow gap laser welding as a candidate process for the manufacture of nuclear pressure vessels. Int J Press Vessels Pip 157:43–50
11. Shi H, Zhang K, Xu ZY et al (2014) Applying statistical models optimize the process of multi-pass narrow-gap laser welding with filler wire. Int J Adv Manuf Technol 75(1):279–291
12. Elmesalamy A, Francis JA, Li L (2014) A comparison of residual stresses in multi pass narrow gap laser welds and gas-tungsten arc welds in AISI 316L stainless steel. Int J Press Vessels Pip 113:49–59

13. Guo W, Li L, Dong SY et al (2017) Comparison of microstructure and mechanical properties of ultra-narrow gap laser and gas-metal-arc welded S960 high strength steel. Opt Lasers Eng 91:1–15
14. Ning J, Zhang LJ, Yang JN et al (2019) Characteristics of multi-pass narrow-gap laser welding of D406A ultra-high strength steel. J Mater Process Technol 270:168–181
15. Katayama S, Abe Y, Ido R et al (2011) Deep penetration welding with high power disk lasers in low vacuum. ICALEO 1:669–678
16. Jones LP, Aubert P, Avilov V et al (2003) Towards advanced welding methods for the ITER vacuum vessel sectors. Fusion Eng Des 69(1/2/3/4):215–220
17. Huang Z, Cai Y, Mu WD et al (2018) Effects of laser energy allocation on weld formation of 9%Ni steel made by narrow gap laser welding filled with nickel based alloy. J Laser Appl 30(3):032013
18. Wu SK, Zou JL, Xiao RS et al (2016) Ultra-narrow-groove laser welding for heavy sections in ITER. Weld J 95(8):300S-308S
19. Zhao Y, Ma SC, Huang J et al (2017) Narrow-gap laser welding using filler wire of thick steel plates. Int J Adv Manuf Technol 93(5):2955–2962
20. Jokinen T, Kujanpää V (2003) High power Nd: YAG laser welding in manufacturing of vacuum vessel of fusion reactor. Fusion Eng Des 69(1/2/3/4):349–353
21. Prasad K, Dwivedi DK (2008) Some investigations on microstructure and mechanical properties of submerged arc welded HSLA steel joints. Int J Adv Manuf Technol 36(5):475–483
22. Vollertsen F, GrÜnenwald S, Rethmeier M et al (2010) Welding thick steel plates with fibre lasers and GMAW. Weld World 54(3):R62–R70
23. Tseng KH, Hsu CY (2011) Performance of activated TIG process in austenitic stainless steel welds. J Mater Process Technol 211(3):503–512
24. Li RY, Wang TJ, Wang CM et al (2014) A study of narrow gap laser welding for thick plates using the multi-layer and multi-pass method. Opt Laser Technol 64:172–183

Chapter 6
Underwater Welding

6.1 Characteristics and Methods of Underwater Welding

With the rapid development of marine engineering, various marine engineering structures such as offshore oil production platforms, subsea oil and gas pipelines, subsea oil depots, subsea tunnels, and offshore airports have been designed and constructed. Ocean engineering structures, due to their year-round work at sea, have extremely harsh working environments. In addition to being subjected to the working loads of the structure, they also have to withstand additional loads caused by storms, waves, tidal currents, as well as seawater corrosion, sand flow erosion, earthquakes, or ice currents in cold regions [1]. Therefore, there are strict quality requirements for the design, manufacturing, material selection, and welding construction of offshore engineering structures in industry.

The main part of marine engineering structures is underwater, and a large number of welding constructions need to be carried out underwater or on the surface of the water. Underwater welding is also a necessary technician for emergency repair of ships, maritime rescue, bridge erection, and other work [2]. Due to the difficulty and high cost of inspecting and repairing underwater welded joints, significant structural damage or overturning accidents can result in serious loss of life and property [1]. Therefore, underwater welding technology is receiving increasing attention from around the world and has made rapid progress.

With the development of the oil and natural gas industry, offshore pipeline engineering is increasingly advancing towards the deep sea. A large number of marine engineering constructions have put forward new requirements for underwater welding technology. Driven by advanced science and technology, underwater welding technology has entered a new stage of development and mainly presents two characteristics. Firstly, based on the needs of social and economic development, a large number of simple and economical underwater welding technologies have been developed, mainly manifested in the expansion of the use of underwater wet welding

R. Li and T. Li, *Advanced Welding Methods and Equipment*,
https://doi.org/10.1007/978-981-97-4109-0_6

technology. The second is based on the development of advanced scientific technology, underwater welding has been more mechanized, automated and intelligent [3].

6.1.1 Characteristics of Underwater Welding

The metallurgical process of underwater welding exhibits different characteristics from land welding due to the influence of water and pressure. The poor visibility of underwater welding, coupled with the influence of bubbles generated around the arc, seriously hinders the normal operation of the welding process. The heat conductivity coefficient of seawater is high, about 20 times that of air. During wet welding, seawater has an obvious quenching effect on the weld, which is prone to high hardness quenching hard tissue. The cooling process will vary greatly due to different welding methods, so that the metallographic structure and mechanical properties of the weld will also vary greatly. Because the welding process is carried out in water, the hydrogen content of underwater welds is generally high. Due to the changes in the composition of the arc atmosphere, hydrogen and oxygen have a particularly significant impact on the welding metallurgical process, which is easy to cause hydrogen embrittlement or other defects such as white spots, brittle fractures, spots and cold cracks. The metallic hydrogen content of the weld seam has now become an important indicator for evaluating underwater welding methods. Due to the influence of deep-water pressure, as the pressure (water depth) increases, the arc voltage increases, splashes and smoke increase, and the content of gases (such as N, O, H, etc.) in the weld metal generally shows an increasing trend. The burning of alloy elements in the weld seam intensifies, and the formation of the weld seam also deteriorates with the increase of pressure. During dry and local dry welding, the arc is not directly affected by the environment such as water and bubbles, but the effect of environmental pressure still exhibits the same characteristics as wet welding arc.

Usually, underwater dry welding requires a long preparation time, and the equipment includes a special chamber with a power supply, a life support system, a pollutant removal system, etc., weighing up to 40 tons. Due to its low operability, high cost, and significant limitations on the welding area and environment, it is mainly used in high-rise structures such as large underwater pipelines and deep-water platforms. Underwater local dry welding uses mechanical means to drain the water from the construction area, isolate the arc burning area and welding area from the water, and form stable gas around the welding area [4]. The submersible welding machine and welding object are in water, thus avoiding the influence of water during the welding process. However, due to the high safety requirements of equipment in water, equipment maintenance costs are high.

6.1.2 Classification of Underwater Welding

In the early eighteenth century, British scientist David discovered that welding arcs could come into contact with water and burn in water, thus opening up a new era of underwater welding. According to this phenomenon, scientists have successfully invented and realized underwater wet arc welding. The principle of underwater wet welding is similar to that of submerged arc welding. When the welding rod comes into contact with the weldment, the surrounding water vaporizes under the action of heat, forming a gas phase zone [5]. When the welding wire leaves the weldment, the arc can be ignited in the gas phase zone mentioned above. Afterwards, under the action of arc heat, more water around the weldment undergoes vaporization and decomposition. At the same time, after the combustion reaction of the coating, gas is generated, which, together with the gas generated by water decomposition, forms bubbles of a certain size around the arc. Bubbles discharge the water around the welding arc, allowing the arc to maintain combustion in the anhydrous gas phase. When the bubbles grow to a certain extent, they will rupture, and new bubbles will constantly be generated during the welding process. This reciprocating makes the welding arc always in the anhydrous gas phase zone, and as the welding arc continues to burn, underwater welding is achieved.

At present, there are many kinds of underwater welding methods being applied and studied in the world, and arc welding is more mature [1]. Underwater arc welding is a kind of arc welding process operation carried out under special conditions (underwater), which is mainly determined by the nature of the environment in which the process is carried out. If the welding is carried out directly in water, it is called wet welding. All accompanying phenomena of the arc depend to a large extent on the physical properties and pressure of the water and the physical chemical process under the influence of the arc on the water environment. If the welding area is separated from water and the welding process is in a dry or locally dry environment, it is called dry welding or local dry welding. The welding arc is related to the composition and pressure of the surrounding gas.

With the development of underwater welding technology, in addition to the commonly used underwater wet welding, underwater local dry welding and underwater dry welding, there are also some new underwater welding methods, such as underwater laser welding, underwater explosion welding, underwater friction stir welding, underwater submerged arc welding [1]. Generally, underwater welding methods are divided into three categories, namely wet method, dry method, and local dry method underwater welding [6]. We will introduce underwater welding according to the process shown in Fig. 6.1.

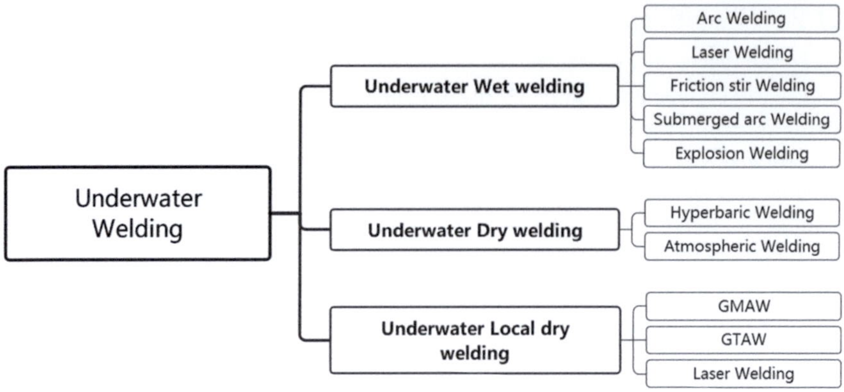

Fig. 6.1 Classification of underwater welding

6.2 Underwater Wet Welding

Wet underwater welding, as the name suggests, refers to the welding process conducted in underwater environments. The principle is shown in Fig. 6.2. Compared to traditional dry welding, underwater wet welding involves dealing with challenges such as water pressure, temperature, and flow, requiring the use of specialized equipment and techniques to ensure the feasibility and safety of the welding process.

During underwater wet welding, welders typically wear specially designed diving suits or full-body diving gear to ensure their safety and comfort underwater as shown in Fig. 6.3 [7]. The welding equipment and power sources must also be specially designed to prevent water ingress and corrosion. In the underwater wet welding

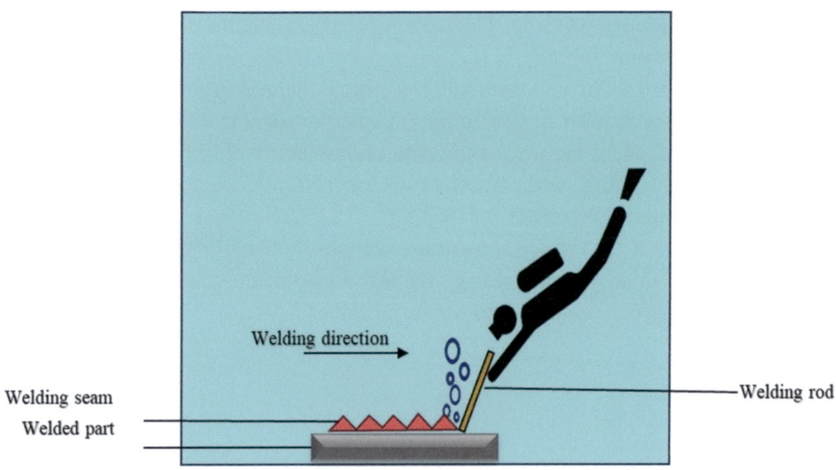

Fig. 6.2 Schematic diagram of underwater wet welding

Fig. 6.3 Actual operation of underwater wet welding [7]

process, welders use specially made welding electrodes and welding materials to carry out the welding operations. The welding arc and current generate enough heat to melt the welding material and join them together. Bubble curtains or water sprayers may be used to isolate the welding seam from water contact.

Wet underwater welding has significant applications in underwater engineering, marine structure repair, and maintenance, among other fields. However, due to the unique and challenging operating environment, underwater wet welding requires highly skilled welding techniques and experience, and strict adherence to safety procedures to ensure welding quality and welder's safety.

6.2.1 Underwater Wet Arc Welding

(1) Principle of underwater wet arc welding

Underwater wet arc welding is a process where divers and welders directly weld in water, without a mechanical barrier between the water and the welding arc. The operation process is simple and feasible. For geometrically complex structures, welding can be carried out, but water can have a serious impact on the quality of the weld seam. Underwater wet welding is widely used in shallow water areas with good ocean conditions and for welding components that do not require high stress. The most widely used area for underwater wet arc welding internationally is the Gulf of Mexico.

A necessary condition for the existence of underwater arcs is the presence of steam bubbles formed around the arc. It occurs at the beginning of welding. When the electrode contacts the workpiece due to the action of current, heat is released.

When the heat accumulates enough to form water vapor, a bubble will appear around the contact point, stimulating the arc within the interval formed between the electrode and the workpiece. The underwater arc burns in a closed steam bubble, constantly changing and replenishing through the combustion evaporation products, mineral coatings, and hydrolysis of the molten electrode. The formation of steam bubbles around the arc is related to the back pressure of atmospheric pressure and the pressure of the water column above the arc, that is, the pressure in the bubbles is directly proportional to the depth of the arc immersed in water (welding depth).

The underwater wet welding arc has the following characteristics: (1) The periodic rupture of bubbles around the arc will generate a large number of floating small bubbles, which has a certain interference effect on the stability of the arc bubbles; (2) Under the action of water depth pressure, the arc is compressed, the cross-sectional area decreases, the voltage increases, and the static characteristics of the arc show an upward trend; (3) When the water depth pressure increases, the arc length shortens. To maintain a normal arc length, the arc voltage should generally increase by about 1 V for each additional atmospheric pressure.

In underwater wet welding, due to the fact that the heat conductivity of water is much higher than that of air, the solidification speed of underwater molten pool will be much faster than that on land, which will cause the cooling speed of welded joints to be too fast, forming quenched hard tissue in the weld and heat-affected zone, reducing the strength and toughness of welded joints, increasing the diffusible hydrogen content in the deposited metal, and also easily producing cracks, pores, slag inclusion and other defects, seriously affecting the welding quality. Therefore, researchers have been committed to developing new underwater welding methods to address these quality issues. For example, Oliver Brätz [8] et al. used induction heating technology for underwater wet shielded metal arc welding (SMAW) to suppress hydrogen induced cracking in underwater wet welding. The equipment is shown in Fig. 6.4. In addition, the presence of water environment also has complex and uncontrollable effects on bubbles, droplets, arcs, and molten pools, leading to adverse consequences such as excessive splashing and poor weld formation. The underwater welding pores are mainly divided into two types: hydrogen pores and carbon monoxide pores. The formation and growth of underwater welding pores are related to the advancing speed of the solidification interface and the hydrogen content in the liquid layer at the front of the interface, and these pores exhibit different morphological characteristics and formation mechanisms as the water depth increases [9]. For the quenched hard structure and defects of underwater wet welding seams, medium frequency induction heating, tempering weld bead technology, or insulation materials can be applied to the base material to isolate water from the workpiece. These methods can reduce hard and brittle structures such as martensite and upper bainite, refine grains, reduce the cooling rate of the weld, lower the average hardness of the heat affected zone, and effectively improve the cold crack sensitivity of the heat affected zone.

The most commonly used underwater arc wet welding is electrode arc welding and flux cored wire self-protection welding. The cost of electrode arc welding is small. In flux cored wire self-protection welding, adjusting flux core composition can

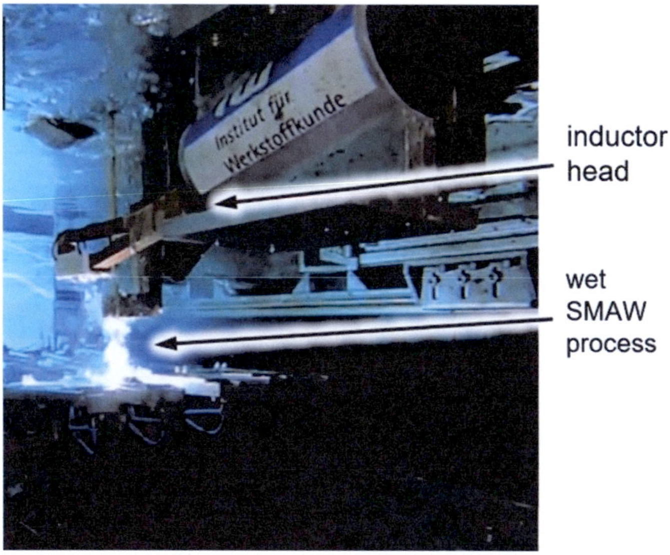

Fig. 6.4 Underwater wet SMAW with induction heating technology [8]

significantly improve the weld quality. In underwater wet welding, researchers mainly focused on the research of coated electrodes in the early stage, and the latest progress mainly focuses on the welding mechanism and performance of cored self-protection welding [10].

(2) Welding materials for wet arc welding

At present, the welding materials for wet arc welding are mainly divided into welding rod and flux cored wire. By adding additives to the welding material to transit it into the weld seam, the microstructure and mechanical properties of the welded joint are improved, and the stability of the welding process is enhanced. The formula of underwater welding rods or flux cored welding wires directly determines the microstructure of the welded joint and has a significant impact on the mechanical properties of the weld. In response to the common problems of porosity, cracks, slag inclusion, and segregation during underwater welding, a series of formula adjustments are required for underwater welding rods or flux cored welding wires to meet the requirements of underwater use and obtain high-quality welded joints:

(1) Having an effective waterproof layer, in order to ensure the stable combustion of underwater welding arcs, it is necessary for underwater welding rods to provide a waterproof effect, as underwater wet welding steel and underwater welding rods need to come into direct contact with water. This can enable the arc to burn in a relatively dry small environment.
(2) Adding hydrogen removal elements to the underwater welding rod coating can eliminate the influence of hydrogen atoms on residual stress in the weld, thereby reducing cracks in the weld and improving the mechanical properties of the weld.

(3) Adding alloying elements such as Cr, Mo, V, Ti to the coating formula of underwater welding rods can strengthen the weld seam. The addition of alloying elements can significantly improve the grain structure of the weld seam, and improve the tensile strength, microhardness, impact strength, and other mechanical properties of underwater welding joints. At the same time, some alloying elements can improve the corrosion resistance of the weld seam, which thus improve the hull, pipeline and the service life of components in the ocean.
(4) According to the different acidity and alkalinity of the oxide coating of underwater welding rods, they are usually divided into acidic welding rods and alkaline welding rods. According to different microstructures, underwater rods can also be divided into ferritic underwater rods and austenitic underwater rods. Iron oxide underwater electrode and rutile underwater electrode are two common ferritic underwater rods [11], and Ni based underwater electrode is a common austenitic underwater electrode.

Common underwater flux cored self-protection wires usually use rutile as the coating material, and are compounded with an appropriate amount of iron powder, nickel powder, aluminum powder, molybdenum powder, titanium powder, boron powder, MgO, ZrO_2, SiO_2, K, Na, Al_2O_3, Si, $CaCO_3$, CaF_2, ferromanganese alloy, aluminum magnesium powder, rare earth elements and other alloying elements to obtain high-performance welded joints with good shape and few defects.

6.2.2 *Underwater Wet Laser Welding*

Underwater wet laser welding has the advantages of fast welding speed, low heat input, small heat affected zone and small deformation. When using fiber laser, the beam can be transmitted to the position to be welded through the fiber for a long distance, without the intervention of diving welders, and it is easy to realize automatic and accurate positioning welding. The most outstanding advantage of underwater laser welding compared with other wet welding technologies is that the laser transmission is not affected by the water depth pressure, but only related to the water penetrated in the transmission process, so it can be welded under high environmental pressure in the deep sea [12]. Underwater laser welding differs from traditional laser welding, as shown in Table 6.1.

Research has shown that water depth has a significant impact on the weld formation and mechanical properties of laser underwater wet welding. The stability of the underwater wet laser welding process and the quality of the weld gradually decrease with the increase of water depth. The scattering and refraction of bubbles in the water, as well as the disturbance of the water surface, cause the laser beam to decay [13]. When the water depth exceeds 200 m, a large number of pores appear on the surface of the weld. When the water depth exceeds 300 m, the plasma generated by laser induction causes a shielding effect on the laser beam. When the water depth exceeds 600 m, the non-uniformity of the weld increases. When the water depth is 700 m,

Table 6.1 Comparison between traditional laser welding and underwater laser welding

	Traditional laser welding	Underwater laser welding
Typical weld microstructure	No water effects	Water effects: increased hydrogen content and metallurgical pores in welds under water; grain refinement; poor weld formation
Weld performance	High	Relatively low
Welding mechanism	Heat transfer mechanism; key hole mechanism	Heat transfer mechanism; key hole mechanism
Welding deformation	Relatively larger	Relatively smaller

the weld seam cannot be fully formed, and the number of pores further increases. In terms of current underwater laser wet welding, although the welding equipment is simple and the cost is low, there are shortcomings in ensuring the welding quality.

6.2.3 Underwater Wet Friction Stir Welding

Underwater wet friction stir welding uses water as the cooling medium, which is the most typical type of immersed friction stir welding. Water can reduce the peak welding temperature, achieve the effect of forced cooling of welded joints, shorten the high temperature residence time, and reduce oxidation inclusions in the joints, as shown in Fig. 6.5. In the process of friction stir welding, flowing water is applied on the surface of the workpiece, and the tensile strength of the welded joint obtained after welding is significantly higher than that of the ordinary joint due to the cooling effect of water, which indicates that underwater friction stir welding can effectively improve the mechanical properties of the ordinary joint. For the friction stir welding process without molten pool and arc, the influence of water environment on the welding process in a certain water depth is mainly reflected in the change of heat dissipation characteristics of the workpiece.

When the underwater friction stir welding technology is used to weld aluminum alloy, fine equiaxed crystals are formed in the weld nugget zone through dynamic recrystallization, and the cooling effect of water makes most of the precipitates dissolve into the aluminum matrix. Compared with the conventional friction stir welded joint in air, the microstructure of the weld nugget zone of the water-cooled joint is finer, the precipitation phase in the heat-affected zone of the weld is of more amounts and smaller, and the weld softening zone is narrower. The tensile strength of underwater welded joints is significantly improved compared to those welded in air.

At present, the research on underwater friction stir welding mainly includes the following aspects:

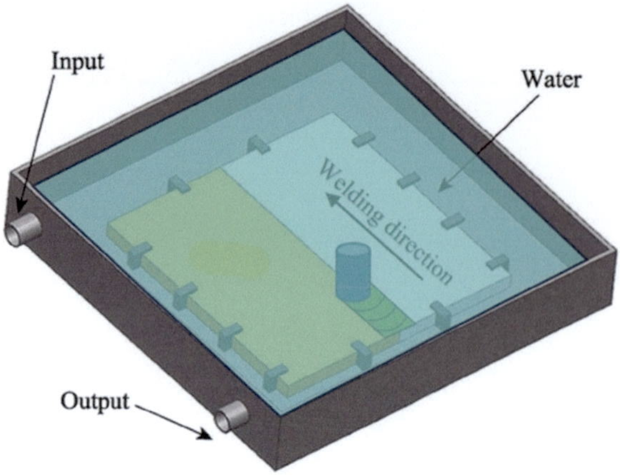

Fig. 6.5 Schematic diagram of underwater wet friction stirs welding

(1) The external design of the stirring head shoulder and stirring needle and their impact on the weld performance;
(2) The effect of welding parameters on weld performance;
(3) Action mechanism of force and heat in friction stir welding;
(4) Weldability of dissimilar materials.

Although underwater friction stir welding can solve some material connection problems that cannot be welded by fusion, the current research on underwater friction stir welding mainly focuses on improving the mechanical properties of aluminum magnesium alloy working members, while the improvement effect on underwater steel members is still far inferior to the relevant underwater fusion welding. In addition, due to the complexity of underwater friction welding mechanisms and the large volume of equipment, their application in underwater engineering has certain limitations. At present, research on underwater welding mainly focuses on correcting the thermal cycle of friction stir welding.

6.2.4 Underwater Wet Submerged Arc Welding

Underwater wet submerged arc welding is a widely used welding method in industrial production, which has the advantages of high welding efficiency, large penetration depth, and good welding quality. The principle is shown in Fig. 6.6. The underwater wet submerged arc welding flux has a protective effect on the welding process. On the one hand, the flux is difficult for water to enter the welding area, and on the other hand, the metallurgical treatment effect of submerged arc welding flux will further improve the welding quality. Since the twenty-first century, efficient welding has always been an important goal pursued in the welding field. In order to further improve the efficiency of submerged arc welding, a large number of scholars have developed multi wire submerged arc welding, strip electrode submerged arc

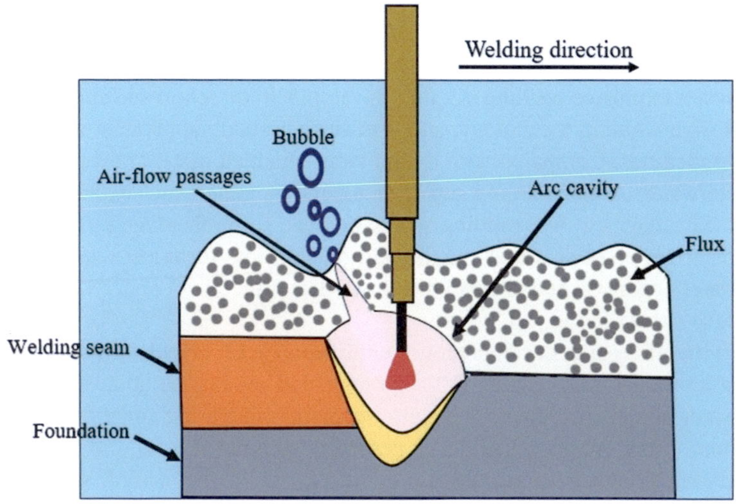

Fig. 6.6 Schematic diagram of underwater submerged arc welding

welding, etc., by changing the number or shape of electrodes; alternatively, narrow gap submerged arc welding was developed by changing the shape and size of the groove; in addition, there are also cold wire and hot wire submerged arc welding [14].

Underwater submerged arc welding technology is based on underwater wet welding technology and draws inspiration from the idea of onshore submerged arc welding. It is a method of welding by applying a mixture of different components of welding consumables to the welding area. One of the core principles of this technology is to use mixed solder to protect the welding area and prevent water intrusion. The second is to use the flux components in the solder to form thick and dense slag to protect the weld seam after welding, thereby reducing the cooling rate of the welded joint. Therefore, the composition of the solder is the key to the success of this technology.

Water glass can be used as a solvent material and mixed evenly with melted submerged arc welding flux before being applied to the area for welding. The addition of mixed solder can significantly reduce the frequency of arc extinguishing and short circuit during the welding process, maintaining a stable welding process. Due to the insulation effect of the mixed solder, the cooling speed of underwater submerged arc welding joints is significantly reduced. The addition of water glass effectively improves the weld formation of underwater submerged arc welding, and the comprehensive mechanical properties of the welded joints are significantly improved.

6.2.5 *Underwater Explosive Welding*

Underwater explosive welding is a new development direction and application field of explosive welding technology, and it is an important supplement to underwater welding methods. Compared with the previous shielded metal arc welding method for underwater connection and repair of pipelines, underwater explosive welding has the advantages of fast welding speed, no need for preheating, post heating and other heat treatment processes, high weld quality, low technical requirements for underwater operators, and low welding costs. The principle of underwater explosive welding is shown in Fig. 6.7.

With the increasing development of continental shelf resources and ocean engineering construction, the joining technology and repair operation technology of underwater oil and gas pipelines are also receiving increasing attention. At present, the joining of submarine pipelines is generally completed using ordinary welding methods on the water surface before being placed on the seabed, which is relatively difficult in deep-sea operations; and the damaged pipeline is usually repaired by shielded metal arc welding, which has poor welding quality and is difficult to operate. If the explosive welding method can be introduced to underwater pipeline joining, huge economic benefits will be obtained.

6.3 Underwater Dry Welding

When conducting underwater dry welding, it is necessary to design and manufacture complex pressure chambers or workrooms. The principle is shown in Fig. 6.8. According to the different internal pressures in the pressure chamber or studio, underwater dry welding can be divided into atmospheric dry welding and hyperbaric dry welding. Hyperbaric dry welding involves manually setting embankments in shallow water to block the water around the welding area, or using high-strength enclosed pressure chambers to wrap the area to be welded in deep water and maintain the indoor pressure at standard atmospheric pressure, creating an environment similar to onshore welding and achieving high-quality welding results. However, this method is too costly, difficult to implement, and has developed slowly so far. Atmospheric

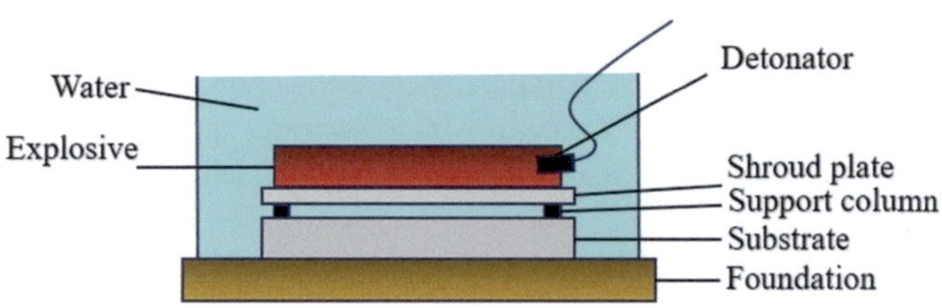

Fig. 6.7 Schematic diagram of underwater explosive welding device

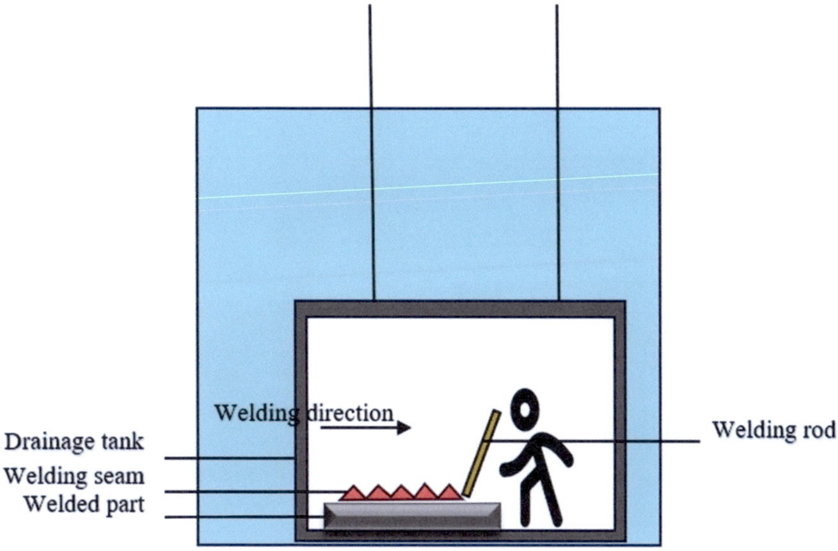

Fig. 6.8 Schematic diagram of underwater dry welding

pressure dry welding requires a specific drainage chamber to isolate water from the workpiece to be welded, and the pressure inside the drainage chamber is consistent with the water pressure at the position of the drainage chamber. The pressure on the chamber casing is relatively low, and the material requirements are relatively simple.

6.3.1 Hyperbaric Underwater Dry Welding

The underwater welding methods commonly used in the hyperbaric chamber at present include gas tungsten arc welding, manual metal arc welding, consumable inert gas welding, and flux cored wire arc welding. Yu [15] simulated underwater dry hyperbaric welding by the flux-cored arc welding (FCAW) method. The underwater dry hyperbaric FCAW system is illustrated in Fig. 6.9. It is possible to monitor and the welding process and collect the arc information by the high-speed camera equipment. In addition, a robotic arm or a three degree of freedom robotic arm that can grasp welding rods can be installed in the cabin, which can be used for all-position linear welding. The movement of the specimen is driven by a driving motor. At the same time, the cabin can also be equipped with a water cooling system, gas circulation cleaning system, smoke removal system, protective gas supply system and necessary electric control system. By adding bottom and top control computers to the hyperbaric test chamber, the control accuracy of the equipment can be effectively improved, which is conducive to the improvement of welding quality. In addition, by designing a remote handle outside the cabin and a walking car, the welding gun

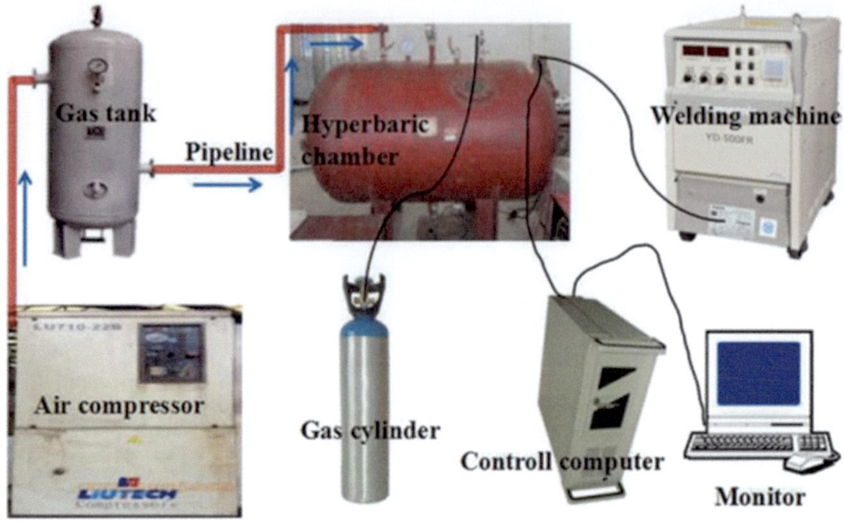

Fig. 6.9 Schematic illustration of the underwater dry hyperbaric FCAW system [15]

inside the cabin can be controlled to move up and down, and the welding process can be monitored through an observation port to achieve automatic welding.

High pressure underwater welding also needs to be used in hyperbaric underwater welding repair operating systems. Hyperbaric underwater welding repair equipment is mainly used for underwater pipeline repair operations. In foreign countries, track GTAW welding is commonly used to complete welding in hyperbaric environments. Some well-known systems include PRS and OOTO [16]. The PRS system developed by Norwegian Statoil Company is expected to complete automatic welding at a depth of one kilometer. The designers have equipped the hyperbaric welding chamber with energy supply and communication facilities, which are fully controlled by the ground control system. The innovation of this welding system is that it achieves symmetrical welding on both sides of the pipeline, reducing welding deformation from top to bottom. The OOTO system also utilizes tracking GTAW welding machines and fiber optic signals to monitor and analyze effective welding processes.

With the development of offshore oil and gas resources, a large number of structural parts such as oil platform and submarine pipelines are used underwater. And these structural components have been working in environments with seawater erosion and water flow impact for a long time. With the increase of service life, the workload of underwater welding repair for these structural components will continue to increase. Hyperbaric dry metal arc welding (GMAW) is considered a practical deep-water welding method due to its excellent weld performance and deep water depth applicability [17].

The operating conditions for underwater welding of submarine pipelines are completely different from those on land. Therefore, in order to ensure the welding quality of underwater pipelines, high-pressure dry welding is required. Underwater high-pressure dry welding, arc combustion, and droplet transfer need to be carried out in a high-pressure gas environment. The environment pressure, humidity, cooling rate, composition of environmental gases, and welding polarity under hyperbaric environment have a significant impact on the microstructure and properties of underwater hyperbaric dry GMAW welding joints. Multiple studies have shown that as environmental pressure increases, the GMAW welding arc shrinks, the arc length gradually shortens, the stiffness decreases, and the arc stability gradually deteriorates.

Cranfield University conducted research on hyperbaric dry GMAW welding within a water depth of 2500 m in the late twentieth century, mainly focusing on the influence of environmental pressure on welding arc, droplet transfer, and welding pool. It has been confirmed that the process of combining pulse current with short arc length within a water depth of 2500 m can ensure the stability of the welding process and achieve satisfactory weld formation.

6.3.2 Atmospheric Underwater Dry Welding

Atmospheric underwater dry welding is carried out in a sealed pressure chamber, where the pressure inside the chamber is equal to the atmospheric pressure on the ground and independent of the environmental water pressure outside the chamber. In fact, this welding method is not affected by the depth of water or the action of water, and the welding process and quality are the same as when welding on land. The cost of atmospheric pressure dry welding equipment is even more expensive than hyperbaric underwater dry welding, and there are more welding auxiliary personnel, so it is generally only used for deep water welding of important structures. The biggest advantage of this method is that it can effectively eliminate the influence of water on the welding process; and its welding conditions are completely the same as when welding on land, so its welding quality is the most guaranteed.

6.4 Underwater Local Dry Welding

Compared with underwater wet welding with simple equipment, as well as underwater dry cabin hyperbaric welding with expensive and high-quality equipment, the economy and quality of underwater local dry welding are between the two. The principle is shown in the Fig. 6.10. Local dry welding uses a small drainage cover to wrap the welding gun and form a local dry space for power arc combustion through inflatable drainage. Since the middle of the twentieth century, underwater local dry welding has begun to be studied and applied [18]. In addition to the repair of shallow marine structures, this method is particularly suitable for the repair of underwater structures in nuclear power plants that are not deep in water and have radioactive contamination because it is easy to achieve automation by using GMAW, etc. At present, diving welders are still used for maintenance in nuclear power plants with

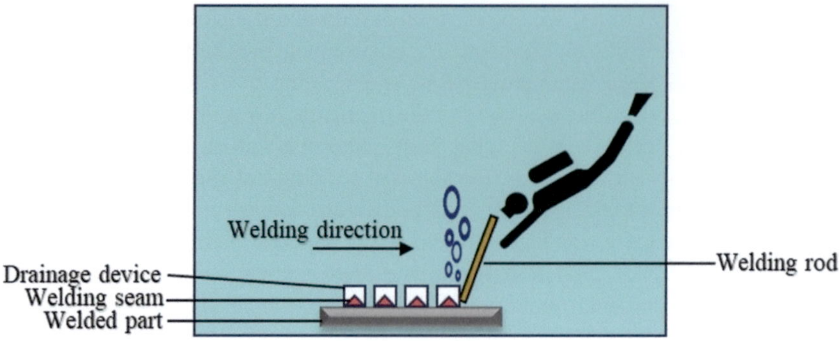

Fig. 6.10 Schematic diagram of partial dry welding

low levels of radioactivity [19], but its development trend is to achieve remote control and automation operations [20], especially in high level radioactive environments. China began researching underwater local dry welding in the 1970s, with the most representative being the semi-automatic CO_2 welding technology for underwater local drainage. This technology has been used multiple times in underwater construction of marine engineering since 1979. China's nuclear power construction started relatively late, and the construction and maintenance of power stations mainly rely on foreign technical forces. Because of the needs of energy development strategies, China has determined to independently innovate and vigorously develop nuclear power, and the nuclear power export project that has been carried out requires China to be fully responsible for the construction and maintenance of a complete set of technical work for nuclear power plants. Therefore, the research and development of welding and repair technology for underwater structures in nuclear power plants has become an extremely urgent practical problem.

6.4.1 Underwater Local Dry GMAW

The main characteristics of GMAW are the use of molten welding wire as the electrode, the use of external gas as the arc medium, and the use of protective gas for arc combustion. It has the characteristics of fast wire melting speed, high deposition efficiency, and low smoke and dust. The deformation after the formation of the weld seam is small, and the welding production efficiency is high. Moreover, due to the good stability of GMAW's arc state and smooth droplet transfer, splash generation is minimal, making it very suitable for building local underwater dry welding platforms. When GMAW is equipped with special external nozzles and water baffles, semi-automatic or automatic local underwater dry welding is fully feasible.

At present, the majority of underwater local dry GMAW repair systems have a significant difference in weld quality compared to onshore ones. The application scenario is temporary underwater connections/additive repairs. However, due to the enormous advantages of GMAW technology such as efficiency, quality, and low consumption, many relevant research institutions and enterprises at home and abroad have made various attempts to improve the weld forming quality of local dry underwater GMAW technology. Neptune Marine Services Ltd. in the United States has developed a local dry underwater GMAW welding system NEPSYS in recent years. This system has been able to obtain high-quality underwater welds, which can be used for permanent welding repair connections, and can provide customized repair solutions under low-cost conditions. However, the NEPSYS system still requires skilled divers to carry complex and bulky diving equipment into the water for local dry underwater GMAW welding. In addition to the long training period for operators, the complex underwater conditions pose a certain risk to operators' personal safety, requiring a complete life support system, which additionally increases the cost of underwater welding.

With the rapid development of unmanned underwater vehicles, researchers are committed to combining them with onshore welding equipment and developing automated underwater welding technology to replace manual repair of complex and dangerous underwater environments.

6.4.2 Underwater Local Dry GTAW

In recent years, the national nuclear power development has made remarkable achievements, which is the trend of future world power development and the need of low-carbon economy development. In order to reduce the impact of nuclear radiation, the inspection and maintenance of internal equipment in nuclear power plants often need to be carried out in water [21]. Therefore, studying underwater welding technology suitable for the nuclear power industry is particularly important. Unlike the complex and ever-changing seawater environment and deep-sea pressure faced during underwater welding in the ocean, the main problem of nuclear power underwater welding is how to achieve automated welding and obtain clean, flat, and reliable welds [22]. At present, the main research direction of underwater welding is electrode arc welding and flux cored wire gas shielded welding, but these two welding methods have high welding smoke, low underwater visibility and poor welding quality [23]. GTAW has the advantages of beautiful weld formation, high welding quality, and low welding dust, making it particularly suitable for the construction and repair of underwater nuclear power projects. For underwater GTAW, wet welding is difficult to ensure arc initiation and arc stability during the welding process, while dry welding cannot highlight the flexibility of the GTAW gun [24]. Therefore, the application prospects of GTAW underwater welding mainly focus on local dry methods, and the main problems it faces are how to continuously and effectively drain and seal and create a local dry welding environment.

GTAW can be applied to a variety of welding materials, which are basically not limited by the thickness of the plate. It has good welding adaptability and high welding joint quality. In addition, due to the fact that the tungsten rod itself does not have a droplet transfer process, the interference caused by the change in arc length is less, which facilitates arc combustion and is very suitable as a heat source for underwater welding. The research on underwater local dry GTAW began in the late 1970s. Initially, by improving the drainage device of the wire brush double spray structure, a stable welding process was achieved at a water depth of 200 m, and satisfactory welds were obtained. Underwater Local dry GTAW can be achieved by designing a small shielding hood. The argon gas sprayed from the GTAW pulse gun effectively protects the electrode and molten pool, forming combustion holes. By using protective gas to exhaust the gas below the shielding hood, external water is blocked from entering the interior of the hood, ensuring that the pressure of internal water is in equilibrium with that of external water, and improving the combustion environment of the arc. By designing a coaxial underwater local dry welding process, direct observation of the arc and molten pool can be achieved, and the underwater welding process can be detected, effectively controlling the welding process. While ensuring consistent weld width, the problem of high hydrogen content and fast cooling speed in the weld can be effectively solved, resulting in a weld with high mechanical properties.

6.4.3 Underwater Local Dry Laser Welding

Underwater local dry laser welding technology refers to the manual installation of a device above the part to be welded that can discharge water and the use of air or inert gas as welding protective gas around the welding position, thereby obtaining a certain range of locally stable dry space and reducing the shielding effect of water on the laser [25]. The underwater local dry laser welding method is shown in Fig. 6.11. Compared with the underwater wet laser welding technology, the underwater local dry laser welding technology can not only reduce the harm of moisture in the water environment invading the weld, but also significantly improve the welding quality of the weld, with a very broad application prospect.

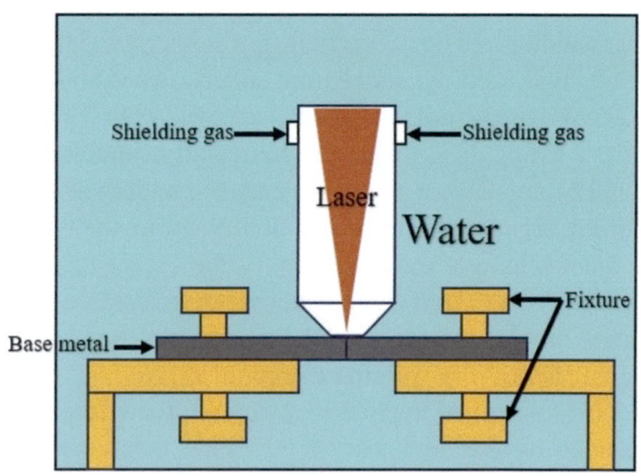

Fig. 6.11 Schematic diagram of local dry underwater Laser beam welding

In 2006, Japanese scholars [26] studied the local dry laser welding repair method of reactor internals for the stress corrosion cracking of reactor internals in the underwater environment of nuclear power plants. When cracks occur [27], it is not necessary to drain the water in the reactor. The welding gun is used to weld the surface where cracks exist on the components. Experiments show that when the base material is austenitic stainless steel, welding with nickel-based alloy filler wire can effectively resist stress corrosion cracking.

6.5 Summary and Prospects

Underwater welding technology is one of the key technologies for developing the ocean, extracting underwater oil, and assembling and repairing large marine structures such as oil extraction platforms, oil pipelines, and underwater warehouses. It is also a necessary technical means for emergency repair of ships, maritime rescue, bridge erection, and other work [28]. Compared to onshore welding, the physical, chemical, and metallurgical processes of underwater welding are extremely harsh, with not only a large amount of hydrogen element present, but also rapid heat dissipation and significant environmental pressure. With the continuous improvement of welding level requirements in engineering projects, it has promoted researchers to conduct in-depth and detailed research on onshore welding technology. With the help of relatively mature onshore welding technology, research on underwater welding technology can be promoted.

(1) Automation technology improvement: In the future, with the progress of automation technology, underwater welding may become more automated and intelligent. Automated robots and drones will be able to perform high-precision and efficient welding operations in underwater environments, reducing reliance on manpower.
(2) Efficient materials and welding methods: The continuous development of new materials and welding methods will make underwater welding more feasible and effective. For example, developing better corrosion resistant materials and welding techniques can be used for the maintenance and construction of marine structures and submarine pipelines.
(3) Application of intelligent sensing technology: In the future, underwater welding will benefit from more advanced sensing technology. Intelligent sensors can help welding engineers monitor parameters such as temperature, pressure, and humidity in real-time during the welding process, as well as detect weld quality and defects, improving the accuracy and reliability of welding.
(4) Virtual reality and augmented reality technology: Introducing virtual reality (VR) and augmented reality (AR) technology can provide real-time visualization and simulation of underwater welding processes. This technology can help welding engineers better understand tasks in underwater environments, and provide operational training and troubleshooting.

(5) Environmental protection and sustainability: With increasing awareness of environmental protection and sustainability, underwater welding will focus more on reducing its impact on marine ecology and organisms. Adopting more environmentally friendly welding materials, processes, and equipment, as well as implementing strict environmental standards, will become an important direction for the future development of underwater welding.

These prospects will promote the continuous development of underwater welding technology and make it a greater role in offshore engineering, energy development and marine resources development.

Problems

1. How is underwater welding implemented? What are the commonly used underwater welding techniques?
2. How are environmental factors such as water pressure, hydrodynamics, and corrosion addressed in underwater welding to ensure welding quality?
3. How does underwater welding affect the welding characteristics of different materials such as metals, alloys, and composite materials?
4. How can the integrity and strength of welds be ensured in underwater welding?
5. What are the commonly used welding power sources and electrodes in underwater welding?
6. What advancements have been made in the automation and robotics technology of underwater welding?
7. How is welding quality and defects monitored during underwater welding?
8. What protective measures are taken in underwater welding? How to ensure the safety of welders and equipment?
9. What are the impacts of underwater welding on the marine environment? How to minimize the negative impacts on marine ecology?
10. What are some application examples of underwater welding in marine engineering, such as underwater pipelines, offshore platforms, etc.?
11. What are the advantages and disadvantages of underwater welding compared to traditional welding methods?
12. How are welding parameters selected in underwater welding, such as current, voltage, welding speed, etc.?
13. What research has been conducted on the influence of the heat-affected zone in underwater welding on welding quality?

References

1. Zhu JL, Yu JR, Jiao XD et al (2005) Progress in research and application of underwater welding technology research and application progress. In: Proceedings of the 11th national conference on welding (2). Chinese Welding Society, Harbin, pp 472–475
2. Chen H (2020) Research on interaction mechanism between droplet transfer process of underwater wet FCAW and water environment. Harbin Institute of Technology, Harbin

3. Liu AZ, Wen Y, Huang F et al (2012) New progress in research and application of underwater welding technology. Sci Technol Inf 1:150, 104. (in Chinese)
4. Liang YJ, Xue L, Lü T et al (2009) Underwater welding technology and its application in ocean engineering in our country. MW Metal Forming 4:17–20 (in Chinese)
5. Ye JX (2007) Study on welding-gun inclination detection and wet underwater welding seam-tracking system based on rotating arc sensor. Nanchang University, Nanchang
6. Tang DY, Long B, Zheng SS (2009) The present situation and development of underwater welding technology in offshore oil engineering. MW Metal Forming 4:24–28 (in Chinese)
7. Gonzalez Romero HA, Bastos Blandón EA, Casadiego Miranda LP et al (2023) Influence of heat input on the weldability of ASTM A131 DH36 fillet joints welded by SMAW underwater wet welding. Sustainability 15(14):11222
8. Brätz O, Klett J, Wolf T et al (2022) Induction heating in underwater wet welding—thermal input, microstructure and diffusible hydrogen content. Materials 15(4):1417
9. Liu YB (2012) Underwater wet welding of E40 steel using flux-cored wire. Harbin Institute of Technology, Harbin
10. Huang ZY (2018) Study on underwater laser welding mechanism and key technologies. Tianjin University, Tianjin
11. Yang K (2013) The development and alloying additions of high-strength covered electrode for underwater wet welding. Harbin Institute of Technology, Harbin
12. Qin H, Cai ZH, Zhang P et al (2019) Forming behavior and properties of medium carbon steel by direct underwater laser beam welding. China Surf Eng 32(3):130–137 (in Chinese)
13. Xing X (2017) Design on draining water device and its welding procedure research in underwater laser welding of 304 stainless steel. Harbin Institute of Technology, Harbin
14. Wu CX, Zhang JX, Zhu BK et al (2009) Development and application of high efficiency submerged arc welding technology. Hot Working Technol 38(23):173–177 (in Chinese)
15. Hu Y, Shi YH, Shen XQ et al (2017) Microstructure, pitting corrosion resistance and impact toughness of duplex stainless steel underwater dry hyperbaric flux-cored arc welds. Materials 10(12):1443
16. Liu HB, Chen XQ (2008) Development and application of underwater welding and cutting technology. Dam Saf 1:68–71 (in Chinese)
17. Chen JQ, Jiao XD, Zhou CF et al (2004) Underwater pipeline repair technology and its related problems. Oil Field Equipment 33(1):33–37 (in Chinese)
18. Li K (2014) Research on arc behavior and droplet transfer of dry hyperbaric GMAW. Harbin Institute of Technology, Harbin
19. Tang DY, Niu HL, Xue L et al (2012) Study on underwater hyperbaric dry GMAW. Electric Weld Mach 42(12):1–6 (in Chinese)
20. Jia CB, Zhang Y, Zhao B et al (2016) Visual sensing of the physical process during underwater wet FCAW. Weld J 95(6):202S-209S
21. Liang M, Wang G, Zhong J (2007) Vision-based seam tracking system of the underwater flux-cored arc welding. J Mech Eng 3:148–154 (in Chinese)
22. Lin SY, Song BT, Song TH (1981) A study and application of local dry CO_2 gas shielded semi-automatic underwater welding. Trans China Weld Inst 2(1):12–23, 52–53 (in Chinese)
23. Gao YF, Hu A (2017) Numerical simulation of wind characteristics in local dry welding chamber and its influence on welding arc. Trans China Weld Inst 38(8):59–62, 72+131. (in Chinese)
24. Gao H, Jiao XD, Zhou CF et al (2011) Study on remote control underwater welding technology applied in nuclear power station. Procedia Eng 15:4988–4993
25. Zhou TH, Chen WB, He Y et al (2010) Beam spatial distribution of upward laser through sea-air interface. Chin J Lasers 37(8):1978–1982
26. Morita I, Owaki K, Yamaoka H et al (2006) Study of underwater laser welding repair technology. Weld World 50(7):37–43

27. Yurioka N, Horii Y (2006) Recent developments in repair welding technologies in Japan. Sci Technol Weld Joining 11(3):255–264
28. Yu JR, Zhang YL, Jiang LP (2001) Underwater welding technology and its progress. Weld Technol 4:2–4 (in Chinese)

Chapter 7
Development of the Traditional Welding Process

7.1 Cold Metal Transfer Welding

7.1.1 Introduction

Usually, in traditional welding processes like gas metal arc welding (GMAW) and gas tungsten arc welding (GTAW), the heat generated from the torch warms up the plates to be joined and a wire is fed in this torch, and then melts the plates and wire together. During welding, the short-circuiting transfer is also happened which will lead to splashing of metal drops onto the component surface. In addition, the high heat input during short circuit transfer may cause rapid heating up of the components, thereby leading to big distortion and other defects. So, these GTAW, GMAW category welding techniques are not preferred for welding together thin type sheets of metal. Remarkable low volume of input of heat and extraordinarily stable arc make cold metal transfer (CMT) welding become the most preferable alternative to GTAW and GMAW. CMT welding permits to attain optimized results while joining dissimilar metal and alloys, especially Al and steel [1].

CMT is a modified GMAW welding process based on short-circuiting transfer process developed by Fronius of Austria in 2004. CMT provides controlled method of material deposition and low thermal input by incorporating an innovative wire feed system coupled with high-speed digital control. The wire feed rate and the cycle arcing phase are controlled to realize sufficient energy to melt both the base material and a globule of filler wire. There are two main features of the CMT process: one is low short-circuiting current corresponding to a low heat input, another is the short circuit occurrence in a stable controlled manner.

A CMT welding equipment is mainly composed of a CMT power source, a welding gun, and a gas supply and clamping system as shown in Fig. 7.1 [2]. The CMT welding system is an integrated device. The parameters of the CMT welding are composed of the WFS (Wire Feed Speed), welding speed, genuine length of wire, shield gas flow rate and arc length correction. Additionally, WFS is the crucial

R. Li and T. Li, *Advanced Welding Methods and Equipment*,
https://doi.org/10.1007/978-981-97-4109-0_7

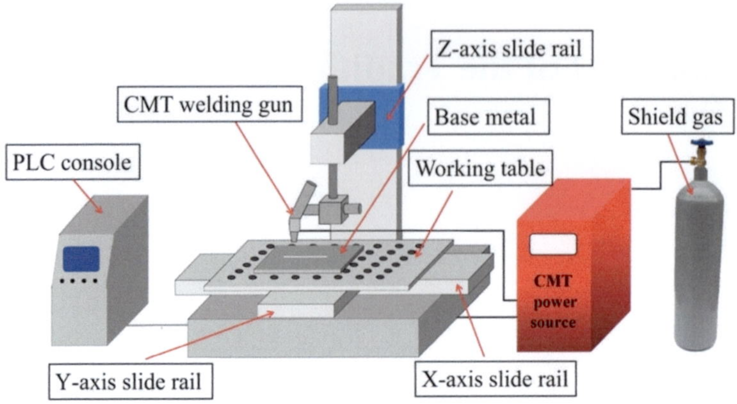

Fig. 7.1 Schematic of a CMT welding equipment [2]

parameter that decides the heat input, which affects the quality of welded joints [3, 4].

7.1.2 Cold Metal Transfer Process

In the CMT process, when the electrode wire tip makes contact with the molten pool, the servomotor of the "robacter drive" welding torch is reversed by digital process control [5]. This causes the wire to retract promoting droplet transfer. During metal transfer, the current drops to near-zero and thereby any spatter generation is avoided. As soon as the metal transfer is completed, the arc is re-ignited and the wire is fed forward once more with set welding current reflowing. A typical CMT welding electrical signal cycle can be defined as the period required to deposit a droplet of molten electrode into the weld pool. The analysis of current and voltage waveform is essential to study the energy distribution of different phases in droplet transfer process. The cycle is divided into three phases as follows:

(1) The peak current phase: This is a constant arc voltage corresponding to a high pulse of current causing the ignition of the welding arc easily and then heats the wire electrode to form droplet.
(2) The background current phase: The phase corresponds to a lower current. The current is decreased to prevent the globular transfer of the little liquid droplet formed on the wire tip. This phase continues until short circuiting occurs
(3) The short-circuiting phase: In this phase, the arc voltage is brought to zero. At the same time, the return signal is provided to the wire feeder which gives the wire a back-drawing force. This phase assists in the liquid fracture and transfer of material into the welding pool [6].

7.1.3 Operation Principle of CMT

The technique of CMT is an altered GMAW methodology, in which the movement of wire and detachment of droplet are related with one another, thereby regulating the input of heat being supplied to the metal sheets to be welded together. During every phase of short circuit, supply of power is interrupted and the movement of the retraction of the wire is regulated; thereby, completely spatter free process is attained together with low input of heat. The torch employed in the CMT process regulates the movement of the filler wire at a mean frequency ranging up to 70 Hz [5, 7, 8]. Principle of operation of CMT welding during the period of acing illustrated by Germinate et al. is shown in the Fig. 7.2. At the beginning, the melted wire containing filler moves towards the pool of joint. Secondly, the wire of filler gets dipped into the pool of joint and the arc has been extinguished. At this duration, the wire is taken a turn around movement, thereby facilitating the detachment of droplet and at this time, the current relevant to short circuit is maintained at small values. After this, the wire of the filler travels in the forward direction and the entire cycle is repeated again.

Another unique feature of the operation principle of CMT welding is that, during joining, the employed control units identifies the short circuit and then guides the detachment of droplet through wire retraction. As a result, the wire then marches forward and is again pulled backwards immediately after the occurrence of short circuit. Due to this, the heat is being generated only by the arc during the arc flaming stage. This short circuit is then regulated and the amount of current is maintained at low values, leading to the transfer of metal, entirely free from spatter. In addition, the length of the arc can be monitored and easily controlled through mechanical means. Therefore, the stability of the arc is perfect, irrespective of the surface to be joined and welding speed, thereby making it possible to employ CMT welding in all position and for any application.

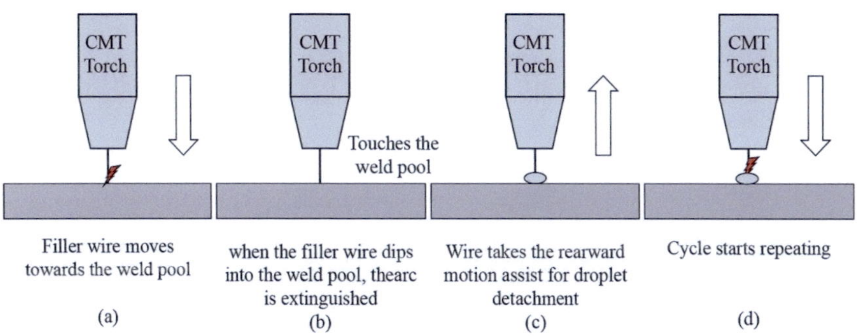

Fig. 7.2 Principle of operation of CMT welding technique as described by Girinath et al. [9]

7.1.4 Characteristics of CMT

The greatest merit of CMT welding is its precision and accuracy. As this joining technique works on both background and foreground based phases in the current, the heat input can be regulated electronically in a perfect manner, which is not possible in the case of GTAW or GMAW welding techniques. Another unique merit of CMT is its high degree of efficiency. For the majority of joining techniques, their consistency relies to some extent on the operator and welder skills; For CMT welding, as it is an entirely electronic based technique, the degree of consistency and efficiency is very high and the chance for error due to human operators is greatly reduced [10].

7.1.5 Applications of CMT

CMT welding is employed to joint together several types of metals including aluminum, inconel, brass, magnesium, stainless steel, etc. The major reason for employing CMT welding for joining thin sheets of metal is that its low input of heat and low volume of voltage is well suited for welding together slender sheets of metal [11]. CMT welding can also be employed to weld together thicker sections of metals, where the soundness of the weldment is of high priority. During the employment of conventional joining techniques, the thicker sections of metals undergo changes in its structural attributes, due to experience of overheating. But when the same metal is joined using CMT welding, the input of heat is restricted to minimum volume, and as a result, the change being encountered in the structure of the metal will be minimum, thereby leading to the attainment of a good quality joint [12]. Moreover, the employment of traditional welding technologies demands a heat safeguarding environment, so as to avoid the wrapping of metal to be welded. But, CMT welding eliminates the need for such shielding. Due to these unique merits, the technique of CMT welding is widely preferred in several industrial sectors involved in the fabrication of marine components, automotive parts, small engine assemblies, joining of pipelines, sheet metals, etc. [13]. In addition, owing to its high degree of accuracy, when compared to other normal GMAW joining techniques, CMT is the most optimal choice for electronic enclosures, joining of automotive panels, etc. [14].

7.1.6 Classification of CMT

Since CMT welding method was invented in 2004, after years of development, CMT welding method has gained a lot of progress and it can be divided into the following categories.

(1) CMT + P

CMT + P (i.e., CMT with pulses addition) is a brand new technique of CMT welding, in which the mode of transfer is a merger of a projected mode of transfer with a single droplet per pulse and a short circuit based mode of transfer during the time of cold metal transfer [15, 16]. In better words, CMT + P welding technique is a merger of cycle of pulses with cycles of CMT (see Fig. 7.3), with escalated input of heat, faster speed of joining, and enhancement in flexibility and performance. In CMT + P welding, an additional droplet is detached in each cycle, thereby leading to larger rates of deposition. During the 1st phase of CMT + P welding, similar to normal CMT process, when the arc is a positive electrode, the retraction of wire takes place. During the phase of pulsed arc, detaching of the droplet occurs simultaneously with the movement of the wire towards the metal plates to be welded. At this point, the arc is devastated and the usual cycle of CMT resumes. Reasonable amounts of experimental work have been carried out with respect to this CMT + P welding technique [17]. Some researchers have proposed a CMT + P welding process involving the use of special units for ultrasonic vibration generation in addition to the conventional CMT machine settings. The ultrasonic category of vibration of the pool of molten metal is brought in by means of a tungsten needle (possessing high point of melting) together with applied ultrasonic type vibration and this needle is inserted from the front portion of the pool, being produced through arc welding. The employed device can generate ultrasonic waves in the range of 20 kHz together with output amplitude of 100 ± 5 μm. The tungsten needle's diameter is around 1.6 mm. Results reveal that the tensile properties of the welded specimen have been improved by the impact of the ultrasonic based vibration and the welded specimen attained by positioning the needle (made out of tungsten) nearer to the pool of molten metal possess highly refined grain structures, together with larger values of tensile related strength.

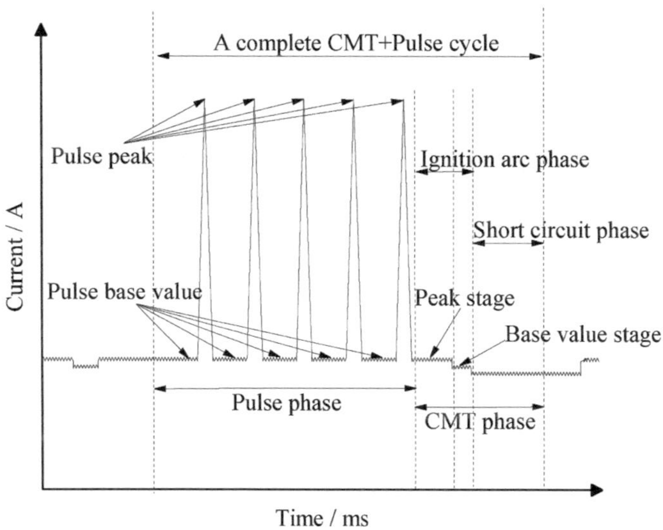

Fig. 7.3 Schematic diagram of the CMT + P welding [18]

(2) CMT advanced welding

The term CMT advanced welding technique refers to the process in which the rates of deposition can be adjusted very precisely by means of negative and positive process cycles. takes place in the short-circuit phase, this joining process ensures the high stability to be expected of cold welding. CMT Advanced also brings further advantages, including targeted heat input, a higher deposition rate with no increase in heat input and minimal distortion. What's more, it produces very few welding fumes, which improves the working conditions for the welder [19–22].

Scotti et al. [23] made an experimental endeavor to figure out the efficiency of the CMT advanced welding technique merged with the technique of near immersion type active based cooling (NIAC) for the wire arc additive fabrication of Al-Mg alloy. Experimental results reveal that the parameters relevant to the electrodes possessing negative and positive polarity electrode (i.e., time share w.r.t. negative and positive polarities) of the CMT advanced welding technique is very significant in regulating the dimension of the layer (i.e., height and width) and waviness of the surface, when compared to that of LEWD (i.e., the distance of water to edge of the layer) based parameters of the technique of NIAC. The efficient width of the fabricated layer gets slender, the height of that layer gets steeper and the waviness of the surface gets coarser, when the positive polarity electrode and negative polarity electrode authorizes lengthier DCEN (direct current negative electrode) fraction of cycle time.

(3) CMT advanced pulse welding

When the positive cycles are alternated with the negative cycles during the CMT advanced welding, CMT advanced pulse welding process combines cycles with a negative electrode polarity and a pulse phase with positive polarity. The process achieves a higher deposition rate during the negatively poled phase with lower heat input. Pole reversal to the positively poled pulsing cycles takes place in the short circuit. As well as higher heat input, the pulse phase gives the user the advantage of non-short-circuiting droplet transfer. The relationship between the positive and negative process cycles is freely selectable. CMT advanced pulse welding process makes it possible to join high-strength steels with a sufficiently high deposition rate yet with low heat input.

In the past few years, the unique merits including slender heat impacted region, minimized distortion, enhanced productivity, etc., have made CMT an best choice welding technique for joining thin as well as thicker sheets and plates of similar and dissimilar metals. In simple words, the technique of CMT welding is an altered gas metal arc type of joining technique, generating less input of heat, larger rates of deposition, highly stable arc and together with complete absence of spattering. In the recent years, many experimental attempts have been put forward to join similar as well as dissimilar metals including superior strength steels, alloys of Al, Cu, Mg, etc., by employing CMT welding.

7.2 Hot-Wire GMAW/GTAW Welding

7.2.1 Hot-Wire GMAW Welding

(1) Introduction

Hot-wire gas metal arc welding (GMAW) welding represents a cutting-edge advancement in the field of welding technology. This section delves into the intricacies of hot-wire GMAW welding, exploring its principles, applications, benefits, challenges, equipment, and its role in pushing the boundaries of welding capabilities.

Hot-wire GMAW welding is an enhanced version of traditional GMAW welding. It involves the introduction of an additional heat source, often an electrically heated wire, alongside the primary arc. This supplementary heat increases the weld pool's energy, allowing for greater penetration and control over the welding process. The arc, in conjunction with the heated wire, creates an environment that facilitates precise control over the weld's characteristics.

The hot-wire GMAW welding system consists of the traditional GMAW welding equipment complemented by a controlled wire heating mechanism. The heated wire is often driven by a separate power source. Advanced control systems synchronize the arc and wire heating to achieve the desired energy distribution and penetration depth. The schematic diagram ofhot-wire GMAW is shown in Fig. 7.4.

(2) Applications

Hot-wire GMAW welding's applications are particularly relevant in critical industries such as aerospace, power generation, and heavy fabrication. It is especially

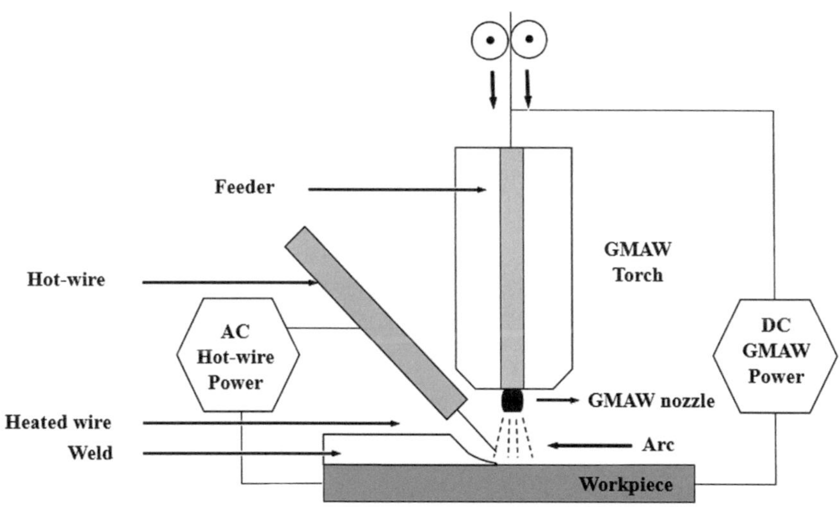

Fig. 7.4 Schematic diagram of Hot-wire GMAW

valuable when welding thick sections and exotic materials, or when demanding exceptional weld quality.

(3) **Benefits and Challenges**

One of the key benefits of Hot-wire GMAW welding is its ability to achieve deep penetration and narrow weld beads without compromising the structural integrity of the joint. Additionally, the controlled heat input minimizes the risk of overheating, reducing the potential for distortion and residual stress. Hot-wire GMAW welding also enables the welding of materials previously considered challenging due to their high melting points or low weldability.

Despite its advantages, hot-wire GMAW welding presents challenges related to process complexity and equipment requirements. The integration of an additional heat source demands precise synchronization between the arc and the heated wire. Proper training and expertise are crucial to harness the full potential of hot-wire GMAW welding and ensure consistent results.

7.2.2 Hot-Wire GTAW Welding

(1) **Introduction**

In the realm of advanced welding techniques, hot-wire gas tugsten arc welding (GTAW) stands as a remarkable innovation. This section delves into the intricacies of hot-wire GTAW welding, exploring its principles, applications, benefits, challenges, equipment, and the pivotal role it plays in achieving precision welds in demanding industries. Hot-wire GTAW welding elevates traditional GTAW welding by integrating an additional heat source, often an electrically heated filler wire, into the process. This supplemental heat augments the primary GTAW arc's energy, resulting in increased weld penetration and controlled fusion. The combined effect of the arc and the heated filler wire allows for meticulous control over the weld pool's behavior, ensuring consistent and precise results.

The hot-wire GTAW welding system involves the integration of a conventional GTAW welding setup with a controlled filler wire heating mechanism. The heated filler wire is often powered independently and synchronized with the GTAW arc using advanced control systems. Proper calibration and synchronization of the two heat sources are crucial for achieving the desired weld characteristics. The schematic diagram of hot-wire GTAW is shown in Fig. 7.5.

(2) **Applications**

Hot-wire GTAW welding finds its niche in applications requiring superior weld quality, intricate geometries, and high precision. Industries such as aerospace, nuclear, and medical device manufacturing benefit from its capabilities. The enhanced fusion and controlled energy input lead to reduced distortion, improved

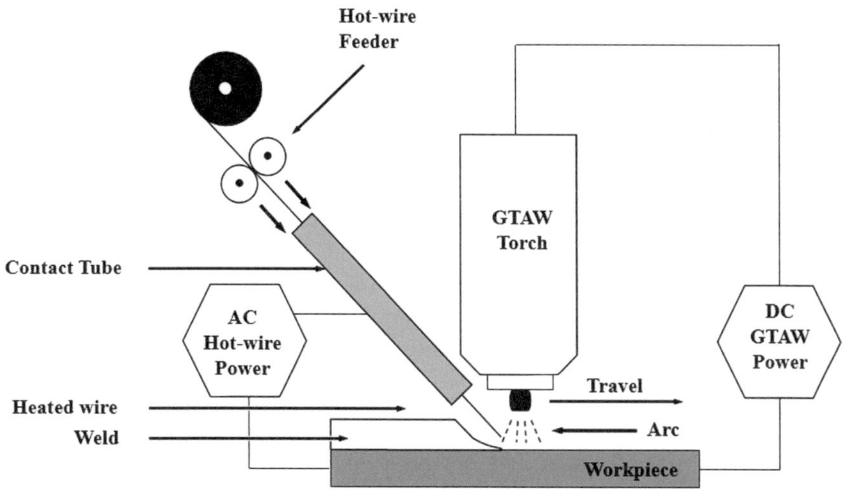

Fig. 7.5 Schematic diagram of Hot-wire GTAW

mechanical properties, and minimized post-weld processing, making it indispensable for critical components.

(3) **Benefits and Challenges**

The primary advantage of Hot-wire GTAW welding lies in its ability to achieve deep weld penetration while maintaining a narrow, controlled weld bead. This combination is ideal for joining materials with varying thicknesses or dissimilar metals. Additionally, the elevated heat input improves the overall metallurgical properties of the weld, enhancing its strength and integrity.

Hot-wire GTAW welding, despite its advantages, presents certain challenges. The complexity of integrating the heated filler wire requires precise synchronization with the GTAW arc. Operator skill and training are vital to optimize process parameters and ensure consistent results. Additionally, selecting appropriate filler wire materials is essential to maintain compatibility with the base materials.

7.3 Tandem and Multi-wire Welding Process

7.3.1 Tandem GMAW Welding

Tandem GMAW is a dual-wire automated GMAW process that increases the application range of traditional single-wire GMAW processes. The process uses a true tandem configuration of two power sources and two wire feeders to generate two separate welding arcs. The output of the two power sources is coordinated by either a robot controller for robotic applications, or by a Tandem GMAW interface for dedicated hard automated work cells. The two arcs are positioned to operate in a common weld puddle by a specially designed high amperage welding torch. Operation of the welding arcs in the close proximity of a common weld puddle is made possible by the use of high-speed waveform control technology to create a special power source output coordinating the interaction of the welding arcs. This unique operating arrangement promotes productivity, often greater than two separate wire electrodes, while providing the operating convenience of a single torch process. Tandem GMAW establishes a means for increasing both the travel speed and deposition rate without compromising the weld quality. The ability of the process to control heat input and bead wetting further enhances weld quality and minimizes weld repair associated with burn-through on thin components, or distortion on thick members.

Dual-wire GMAW systems have only recently become commercially available since 1996. The process offers high deposition rates of up to 24 kg/h at a travel speed of 0.8 m/min as claimed and travel speeds more than twice that of conventional GMAW as reported (up to 5 m/minute for a 2 mm lap joint). At increased travel speeds, heat input is reduced and less distortion occurs. It is also claimed that tandem GMAW reduces porosity because the elongated weld pool extends the degasification time. Metal- or flux-cored wires can also be used to further increase deposition rates. However, this technology can only be used for mechanized or robotic welding because of the precision required in positioning the bulky torch. Joint accessibility is also restricted because of the torch size. The tandem GMAW process has been applied in the automotive industry, and in the shipbuilding, construction equipment, railway vehicle, and boiler construction fields for welding aluminum alloys, steels and other materials. Weld overlays have also been deposited using this technique.

(1) **Process of Tandem GMAW welding**

The Tandem GMAW welding process is shown in Fig. 7.6.

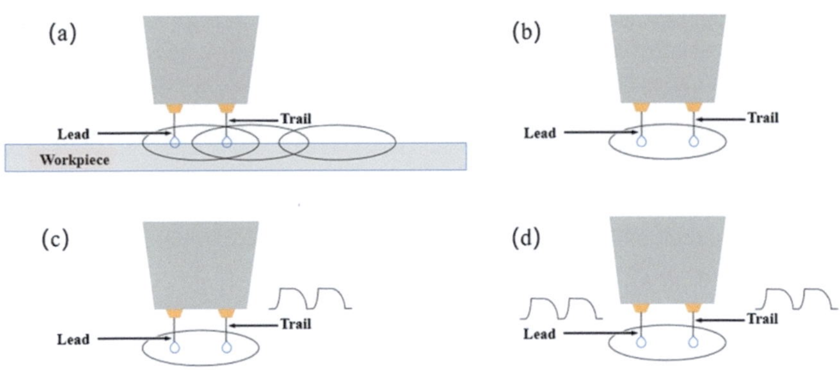

Fig. 7.6 Process of tandem GMAW welding: **a** Two wire electrode in-line configuration; **b** independent arc control with common weld puddle; **c** lead arc in CV mode, trail arc in pulse mode; **d** Synchronized DC pulsed mode

a. Two wire electrodes are positioned in line, in the direction of travel. The first electrode, the "lead", controls deposition rate and penetration. The second electrode, the "trail", controls weld bead appearance.
b. Both electrodes' arcs contribute to a common weld puddle, but each is independently controlled by its respective power source. Depending on application requirements, the lead and trail electrodes may exhibit different operating modes.
c. The lead arc performs in the DC (+) Constant Voltage (CV) mode, while the trail arc operates in the DC (+) pulse mode. Using a CV lead arc maximizes penetration and travel speed. The pulsed trail arc reduces heat input and minimizes electromagnetic arc interference between the two arcs. Additionally, the trail arc cools the molten weld puddle.
d. In this configuration, both the lead and trail arcs operate in a (+) DC pulsed mode. This configuration is typically used to manage total process heat input on thin gauge material and other heat sensitive applications. To use this set-up, the pulse frequency of both electrodes must be synchronized so that the peak of each pulse occurs on the background of the other arc.

(2) **Equipment**

The tandem welding process comprises two completely independent welding circuits, each with its own power source, torch cable, wire drive, and contact tip (see Fig. 7.7). Optionally, a push–pull system is available for aluminum applications.

Tandem welding systems use independent circuits to run the operation. Some provide the option of pulse synchronization between the two power supplies. Pulse synchronization can be used to stabilize the arc by reducing interference between the two welding circuits. Whether pulse synchronization is necessary depends on the specific application. Operators should not assume that pulse synchronization is always necessary. Because of its high deposition rates and intense heat, the tandem welding process is suitable only for automated welding. Therefore, it is important that the welding equipment can be integrated seamlessly with the controller of the automated welding machine. This provides the machine operator/programmer with the flexibility to program the exact welding parameters needed to take advantage of the tandem arc characteristics. If such controls are not available on the machine, then programmable control of the welding parameters should be available on the welding power supplies. Optimum weld quality can be achieved by properly balancing the welding parameters being used by each wire.

The two contact tips are contained within a common torch body, surrounded by a common gas nozzle. The two contact tips are angled in such a way that during welding, the two wires actually share the same arc, and a single molten puddle is formed. Although the tandem welding torch is larger than a standard single-wire torch, equipment suppliers have developed slim, compact torch designs (see Fig. 7.8). This enables good access into many types of welded joints. Generally, the tandem welding process is used in such a way that one welding wire follows directly behind the other in the welding joint, although this is not always necessary. The tandem welding process is effective on carbon steels, stainless steels, and aluminum.

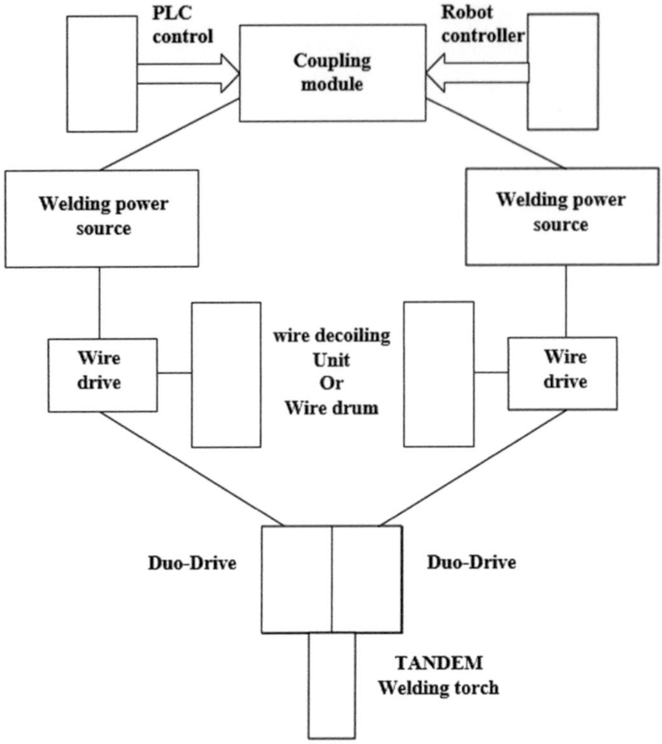

Fig. 7.7 Schematic diagram of tandem welding process

For aluminum applications, push–pull torches are available to ensure reliable wire feeding. Since the lead wire can be set at a high wire feed speed, deep penetration can be achieved, even at high travel speeds. However, complete programmability of both welding wires can also provide limited penetration, if desired, for high-speed welding of sheet metal.

(3) **Advantages**:

i. Exceptional travel speeds: Tandem GMAW often doubles the speed of conventional single-wire
ii. GMAW processes: Depending on application requirements, Tandem GMAW has achieved travel speeds in excess of 200 inches per minute
iii. High weld metal deposition rate: The Tandem GMAW dual wire procedures often exceed 35 pounds of metal per hour
iv. Superior weld quality: Reduced heat input; Resists burn-through on thin materials; Maintains low spatter levels; Produces good penetration on thick materials; Improves bead wetting at weld toes; Resists undercut.

(4) **Differences between dual-wire and tandem welding**

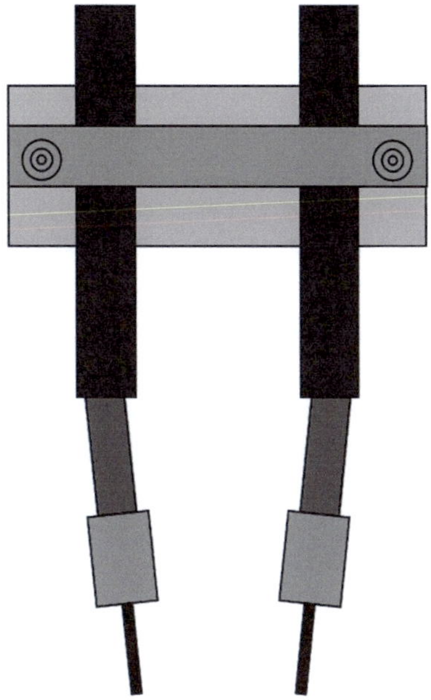

Fig. 7.8 Diagram of tandem welding torch

Double-wire welding, see Fig. 7.9a, represents the very beginning of welding using two wire electrodes. Here, electrodes with the same potential (not electrically isolated) were used. These could either be fed into one single or two separate contact tips. At the time of double-wire welding there were no suitable characteristics such as pulse or PMC (groundbreaking developments). Standard welding in the dip transfer arc used back then led to increased formation of welding spatter, which resulted in instability of the arc and fluctuating arc lengths. This produced other drawbacks in the process (e.g., adhesion of spatter on contact tip and gas nozzle).

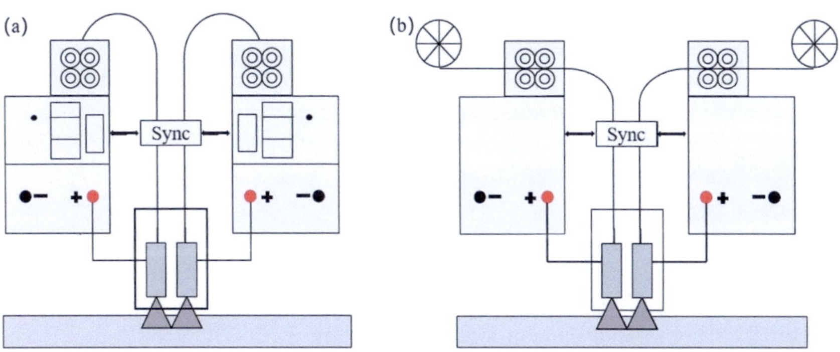

Fig. 7.9 Comparison of double-wire welding and tandem welding: **a** double-wire welding; **b** tandem welding

Tandem welding, see Fig. 7.9b, is based on two independent welding systems that are synchronized with one another. The separate wire electrodes are fed into a single torch hosepack via separate wirefeeders—with a single gas nozzle and electrically isolated contact tips. Two arcs generate a single weld pool. Both wire electrodes can be synchronized accordingly using a twin controller. In addition, a wire electrode can be connected or disconnected. Welding can be performed in both directions. The process is much more stable allowing very high welding speeds (2 m/min +) to be achieved on thin sheets. For thick sheets, the process boasts a high deposition rate (22 kg/h +), which allows a significantly higher weld seam volume to be achieved.

7.3.2 Multi-wire Welding

The recent development in high-productivity GMAW is the use of multi-wire welding systems. As its typical representative, Tandem GMAW welding involves two wires sharing a common gas shield in which the two wires are operated from independent power sources, and it can achieve a deposition rate of 12 kg/h without the flaw of undercut because of a more homogeneous weld pool temperature. To further increase the GMAW efficiency, researchers have developed various triple-wire GMAW processes.

Based on an understanding of the multi-wire welding systems and long-term practical experience, a novel welding method—integrated circular triple-wire GMAW has been developed. In the welding process, three wires are circularly integrated in a big torch and powered by three pulse power sources with cooperative control function respectively. This triple-wire configuration makes the arc heating and welding wire filling more concentrated.

Although triple-wire welding can significantly improve welding efficiency, the interference between the arcs is serious, which may lead to the instability of the welding process. The arc interference is related to the control of pulse currents, and the control of pulse currents depends on the choice of phase angle. At present, the welding of some high degree steel mainly depends on single wire GMAW, which seriously restricts productivity. Therefore, triple-wire welding is expected to achieve efficient and high-quality welding for construction machinery. Heat input is a key influence parameter for joint microstructure and performance. The high heat input by triple-wire welding may lead to the HAZ grain coarsening and deterioration of mechanical properties.

Figure 7.10 shows a schematic diagram of the experimental setup. The experimental device consists of four parts: (1) a welding system, including three Aotai Pulse GMAW-500 welding machines, a special three-wire welding torch, and a welding bench; (2) a high-speed camera system, including a MOTION PRO high-speed camera and a xenon light source [24]; (3) an electrical signal acquisition device, including a NI USB6251 data acquisition card, three LEM Hall current sensor, and three LEM Hall voltage sensors; and (4) a data processing system, including a synchronous trigger device and a computer with data processing software.

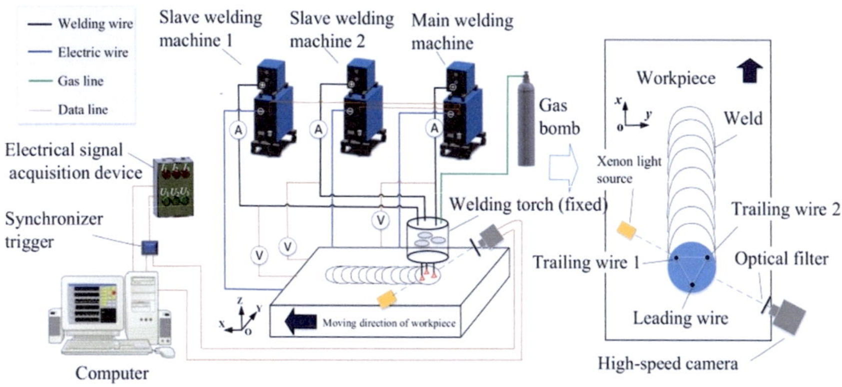

Fig. 7.10 Schematic diagram of the multi wire welding experimental device [24]

The key devices of the welding system are the special welding torch and the three welding machines with communication function. As shown in Fig. 7.10, three welding wires are insulated from each other and circularly arranged with an equilateral triangle configuration in a big welding torch. In order to facilitate the shooting of the welding arc, the welding torch is fixed while the workpiece is moving. Hence, the welding direction is opposite to the moving direction of the workpiece. The welding wire close to the front end of the molten pool is called a guide wire, and the two wires located at the rear end of the molten pool are called trailing wire 1 and trailing wire 2.

The three wires are powered by the three welding machines respectively. All the arcs on the wires are operated in pulse mode, which is conducive to realizeing one pulse per drop of molten droplets (see Fig. 7.11). The cooperative control of the three pulse currents depends on the main welding machine, which sends delay signals to the remaining two welding machines. The other two machines are referred to as slave welding machines. The time relationship of the three pulse currents is represented by the phase angle θ, which is defined as follows:

$$\theta = 360^{deg} \cdot t_\theta / t_{total} \tag{7.1}$$

where t_θ is the delay time of the current pulse between wire 1 and wire 2. Wire 2 and wire 3 adopt the same delay time. t_{total} is the total time of a current pulse cycle.

There are three typical phase angles: 0°, 120°, and 180°. When the phase angle is 0°, the pulse peak phase appears on the three welding wires simultaneously; when the phase angle is 120°, the pulse peak phase appears alternately on the three welding wires, whose state is the same as that shown in Fig. 7.11; when the phase angle is 180°, the pulse phase first appears on the leading wire (wire 2), and after $0.5 \cdot t_{total}$ it appears on the two trailing wires (wire 1 and wire 3) simultaneously.

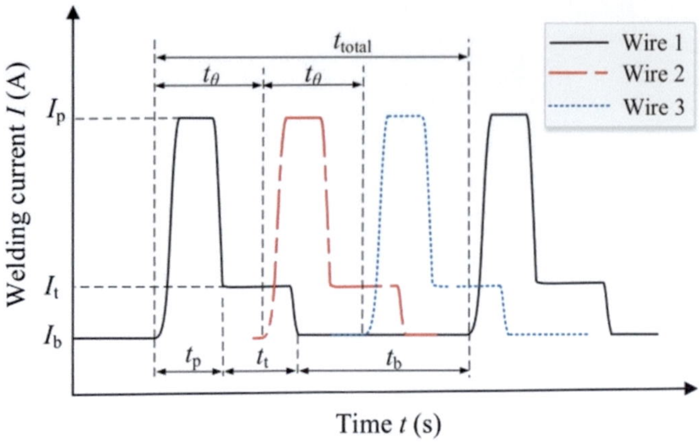

Fig. 7.11 Schematic diagram of welding current waveform for the triple-wire welding [24]

7.4 Plasma Arc Welding

7.4.1 Introduction

Plasma arc welding (PAW) is a type of metal joining process in which coalescence is obtained by heating the work-piece with a constricted arc. The constricted arc can be struck in two ways: One type of arc is generated between the tungsten electrode and the work-piece, which is called a transfer arc; Anotherarc is struck between the electrode and the constricting nozzle, which is called a non-transfer arc.

A schematic of the process is shown in Fig. 7.12. The shielding is provided by a stream of hot ionized gas that flows from the annular gap between the tungsten electrode and the nozzle. This ionized gas may be supplemented by another stream of gas issuing from the annular gap between the nozzle and the outer casing of the welding gun.

Pressure may or may not be applied between the parts being joined. A filler rod may or may not be used in the process. For welding metals of low thickness, it can be done without fillers. For higher section thicknesses, use of fillers is imperative.

7.4.2 Plasma Arc Welding Equipment

The following paragraphs discuss the power source, welding torch, control console and wire feeder used in plasma arc welding.

(1) **Power Source**

Typically, a DC power source is used for plasma welding that has a drooping characteristic and an open-circuit voltage (OCV) higher than 70 V. The OCV is

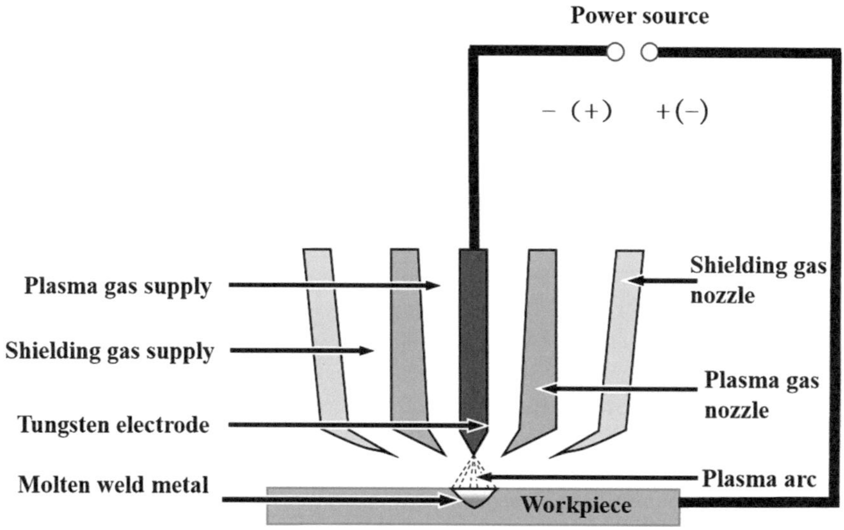

Fig. 7.12 Schematic of plasma arc welding

necessary to initiate and maintain an arc. But, if you use a helium shielding gas, the OCV must be higher than 70 V. Additionally, it is preferable for the power source to have a high-duty cycle if it's necessary to achieve high amperage and long welds. Otherwise, the power source will overheat and require cooling down.

(2) **Welding Torch**

The plasma welding torch is quite similar to the TIG welding torch. It uses a tungsten electrode housed within a collet that's threaded to the torch body. However, unlike the TIG torch, the PAW torch keeps the entire tungsten electrode inside the nozzle without sticking out.

Additionally, the torch must be actively water-cooled. The arc is contained within the nozzle and causes high heat to accumulate. If the water cooling is interrupted, the nozzle will melt. The tungsten electrode used for a plasma arc welding process is EWTh-2, otherwise known as a 2% thoriated tungsten electrode. The nozzle is made of copper, but the plasma nozzle bore diameter must not be too small for the current output and plasma gas flow rate because it can melt and erode fast.

(3) **Control Console**

The control console contains a high-frequency arc starting unit, a non-transferred pilot arc power source, a torch protection circuit, water, and gas valves, delay timing systems, and separate flow meters for the plasma gas and the shielding gas.

(4) **Wire Feeder**

In manual PAW welding, the filler rod is manually fed. In machine welding or automatic welding, the wire feeder can be added to supply a filler metal automatically.

A wire feeder needs to have an adjustable wire feeding speed and have a constant output once set.

7.4.3 Advantages of Plasma Arc Welding

(1) High quality welds can be obtained with this welding method.
(2) High speed welding: The high welding speed of plasma arc welding (PAW) makes it an efficient and productive choice for welding applications, especially those that require large amounts of welding or involve large-scale projects. The constricted plasma arc of PAW results in a more focused and concentrated heat source, which allows for faster and more efficient welding. This increased speed makes PAW a suitable option for high-volume production and other large-scale welding projects where speed and efficiency are important considerations.
(3) The deep penetration capability of plasma arc welding (PAW) is one of its key advantages. This is achieved through the high-temperature and focused ionized gas jet of the plasma arc, which results in a deep and narrow weld bead. This is particularly useful in welding thick or multi-layer materials, where a deep and consistent penetration is essential for ensuring the strength and reliability of the weld. By producing deep and narrow weld beads, PAW can produce high-quality welds that meet the demands of even the most challenging welding applications.
(4) The versatility of Plasma Arc Welding (PAW) has made it a go-to option for many welding projects. PAW's ability to work on an array of metal types, including aluminum, stainless steel, and nickel alloys, among others, has made it an ideal choice for welding, cutting, and spraying operations in various industries and manufacturing processes. With its flexibility and ability to handle delicate and complex components, PAW has proven to be a reliable solution for a wide range of welding applications.
(5) PAW is known for its ability to maintain the original shape of the metal being welded, resulting in minimal distortion. This makes it a top choice for industries where dimensional accuracy is crucial, such as aerospace, precision engineering, or construction. The low distortion capability of PAW provides a welding solution that ensures the metal maintains its original form, making it a highly desired option in these industries where the finished product needs to be of a specific size and shape.
(6) The improved efficiency advantage of PAW refers to its ability to minimize the number of welding passes required to complete a joint, thus leading to a more efficient and streamlined welding process. With PAW, the risk of defects and inconsistencies in the weld is reduced, as the welding can be completed in fewer steps. This saves valuable time and resources, making it a cost-effective solution for many welding applications. Additionally, the reduced number of passes also reduces the chances of human error, leading to a higher quality and more consistent final product.

(7) PAW reduces the likelihood of contaminants affecting the quality of the weld through the use of a shielding gas. This protection helps to minimize the occurrence of defects such as porosity in the weld, resulting in a higher quality finished product. By reducing contamination, PAW improves the efficiency and reliability of the welding process.

7.4.4 Disadvantages of Plasma Arc Welding

(1) The disadvantage of high equipment cost in plasma arc welding (PAW) refers to the fact that it requires specialized and advanced welding equipment, which can be costly to purchase and maintain. The equipment used in PAW is designed to meet the specific requirements of this welding process, making it more expensive compared to other traditional welding methods. The high cost of the equipment and the need for regular maintenance can make PAW a less attractive option for small-scale operations or those with limited budgets. However, the high efficiency and versatility of PAW may make the investment in equipment cost-effective in the long run for large-scale projects or high-volume production.
(2) The setup process for plasma arc welding (PAW) can be quite complicated, as it requires specific equipment and a carefully controlled environment. This means that it may take longer to prepare for a welding job compared to other welding processes, and can also increase the cost of the overall process. Furthermore, the specialized equipment required for PAW can be expensive to purchase and maintain, which may not be feasible for smaller operations or projects.
(3) The success of plasma arc welding (PAW) also relies heavily on the skill of the operator. The process demands a highly skilled operator to execute the welding task and produce high-quality welds. Lack of operator expertise can result in lower-quality welds and may hinder the success of the PAW process, especially in complex welding applications.
(4) Plasma Arc Welding (PAW) has limitations when it comes to welding thicker materials. It is more suitable for welding thin to medium-thick materials, but may not provide the best results when welding very thick materials. The process may struggle to penetrate the thicker material, leading to a lower-quality weld or potential defects. In these cases, alternative welding processes that are better suited to welding thicker materials may be necessary.
(5) The limitation of welding positions in Plasma Arc Welding (PAW) refers to the fact that it is not ideal for welding in positions other than flat or horizontal. PAW is best suited for welding thin to medium-thick materials and may not be the best choice for welding in tight or awkward positions, such as overhead or vertical. This limitation can affect the efficiency and quality of the weld and may require additional resources to complete the weld, such as a welder who is skilled in welding in these positions or specialized equipment to support the welding process.

7.4.5 Principles of Operation of Plasma Arc Welding

The operation of plasma arc welding is somewhat similar to gas tungsten arc welding. There are tungsten non-consumable electrode and shielding gas involved. The main difference is that in PAW, the tungsten electrode is located inside the constricting nozzle, while in GTAW, it is protruding outside the nozzle. Due to the location of tungsten electrode, the arc emanating from PAW gets focused according to the shape of the nozzle. The arc gets concentrated. At high temperature, the arc turns into plasma. The plasma is not exactly gas, nor is it a liquid (and not a solid obviously). That is why it is sometimes called the fourth state of matter—after solid, liquid, and gas. Since the plasma has been constricted in the shape of a stream, the energy in the plasma is concentrated, and temperature of the plasma is higher than the GTAW arc.

7.5 Summary and Prospects

The development of traditional welding processes has led to the emergence of innovative techniques such as Cold metal transfer welding (CMT), Hot-wire GMAW/GTAW welding, Tandem GMAW welding, multi-wire welding, and Plasma Arc Welding (PAW). These techniques offer various advantages, including reduced heat input, higher welding speeds, improved deposition rates, minimized spattering, and enhanced weld quality. CMT welding has been proven to be effective in joining thin and thick sheets of similar and dissimilar metals. Hot-wire GMAW/GTAW welding employs an additional hot wire to achieve independent adjustment of material and energy inputs. Tandem GMAW welding utilizes a dual-wire automated GMAW process to enhance productivity and weld quality. Multi-wire welding, such as tandem welding, demonstrates the potential for higher deposition rates. PAW utilizes a constricted arc to achieve coalescence in metal joining processes.

With regard to the method of welding processes, we can see the following trends and directions of development:

(1) Further advancements in CMT welding: Continuous research can lead to advancements in CMT welding to address specific industrial applications and broaden its range of applicable materials.
(2) Optimization of hot-wire GMAW/GTAW welding: Efforts can be made to fine-tune the hot-wire welding parameters for different materials to maximize efficiency and weld quality.
(3) Enhanced process control in Tandem GMAW welding: Continued development in process control technology can improve the stability and precision of the Tandem GMAW process, enabling better control of heat input and weld bead appearance.
(4) Multi-wire welding applications: Investing in GTAW and developing triple-wire GMAW processes, like integrated circular triple-wire GMAW, can further increase GMAW efficiency and deposition rates for various materials.

(5) Advancements in Plasma Arc Welding: Continuous research can focus on optimizing plasma arc welding for specific applications, including materials with varying thicknesses and complex geometries.
(6) Automation and robotics integration: Integrating these welding processes with advanced automation and robotics can enhance their efficiency and expand their applications in various industries.
(7) Sustainability and environmental considerations: Future developments should also consider reducing the environmental impact of welding processes by exploring eco-friendly shielding gases, consumables, and energy-efficient power sources.

Overall, ongoing research and technological advancements will continue to shape the welding industry, offering improved welding techniques, higher productivity, and enhanced weld quality for various industrial applications.

Problems

1. What are the drawbacks of traditional welding techniques for slender metal sheets?
2. How does the CMT process differ from GMAW welding?
3. What are the three phases of the CMT process?
4. How does CMT welding provide precision, accuracy, and consistency?
5. What are the applications of CMT welding in different industries?
6. What are the categories and variations of CMT welding?
7. How does CMT advanced welding differ from CMT pulse advanced welding?
8. What are the advantages of CMT welding, such as reduced distortion and increased productivity?
9. How has CMT welding been applied in joining dissimilar metals?
10. How does Hot-wire GMAW welding improve economic efficiency by independently adjusting material and energy inputs?
11. What are the limitations and challenges associated with the use of TANDEM GMAW welding, such as joint accessibility and equipment integration?
12. What factors influence the interference between arcs in triple-wire welding, and how can it be effectively managed to ensure welding process stability?
13. What are the advantages of using a water-cooled welding torch in plasma arc welding?
14. What are the characteristics and parameters of the equipment and materials used in Hot-wire GMAW welding experiments?
15. How does TANDEM GMAW welding achieve higher productivity and weld quality compared to single-wire GMAW processes?

References

1. Srinivasan D, Sevvel P, John Solomon I et al (2022) A review on cold metal transfer (CMT) technology of welding. Mater Today: Proc 64:108–115
2. Zhao YL, Chen FR, Cao SL et al (2022) Effect of CMT welding heat input on microstructure and properties of 2A14 aluminum alloy joint. Metals 12(12):2100
3. Tomków J, Sobota K, Krajewski S (2020) Influence of tack welds distribution and welding sequence on the angular distortion of TIG welded joint. Facta Univ, Ser: Mech Eng 18:611–621
4. Bellet M, Cerri O, Bobadilla M et al (2009) Modeling hot tearing during solidification of steels: assessment and improvement of macroscopic criteria through the analysis of two experimental tests. Metall Mater Trans A 40(11):2705–2717
5. Selvi S, Vishvaksenan A, Rajasekar E (2018) Cold metal transfer (CMT) technology—an overview. Defence Technol 14(1):28–44
6. Furukawa K (2006) New CMT arc welding process–welding of steel to aluminium dissimilar metals and welding of super-thin aluminium sheets. Weld Int 20(6):440–445
7. Kannan AR, Shanmugam NS, Vendan SA (2019) Effect of cold metal transfer process parameters on microstructural evolution and mechanical properties of AISI 316L tailor welded blanks. Int J Adv Manuf Technol 103(9):4265–4282
8. Dhanesh Babu SD, Sevvel P, Kumar RS (2020) Simulation of heat transfer and analysis of impact of tool pin geometry and tool speed during friction stir welding of AZ80A Mg alloy plates. J Mech Sci Technol 34(10):4239–4250
9. Girinath B, Shanmugam NS, Sathiyanarayanan C (2020) Studies on influence of torch orientation on microstructure, mechanical properties and formability of AA5052 CMT welded blanks. Arch Civ Mech Eng 20(1):15
10. Eder T (2010) The CMT process—news and its advantages in industry. In: SAE Technical Paper Series. 400 Commonwealth Drive. SAE International, Warrendale, PA, United States
11. Wang P, Hu SS, Shen JQ et al (2016) Effects of electrode positive/negative ratio on microstructure and mechanical properties of Mg/Al dissimilar variable polarity cold metal transfer welded joints. Mater Sci Eng, A 652:127–135
12. Li JJ, Shen JQ, Hu SS et al (2019) Microstructure and mechanical properties of 6061/7N01 CMT+P joints. J Mater Process Technol 264:134–144
13. Zhang H, Hu SS, Wang ZJ et al (2015) The effect of welding speed on microstructures of cold metal transfer deposited AZ31 magnesium alloy clad. Mater Des 86:894–901
14. Sevvel P, Dhanesh Babu SD, Senthil Kumar R (2020) Peak temperature correlation and temperature distribution during joining of AZ80A Mg alloy by FSW—a numerical and experimental investigation. Strojniški Vestnik—J Mech Eng 66(6):395–407
15. Pickin CG, Young K (2006) Evaluation of cold metal transfer (CMT) process for welding aluminium alloy. Sci Technol Weld Joining 11(5):583–585
16. Sevvel P, Satheesh C, Senthil KR (2020) Generation of regression models and multi-response optimization of friction stir welding technique parameters during the fabrication of AZ80A Mg alloy joints. Trans Can Soc Mech Eng 44(2):311–324
17. Tapiola J (2017) Cold metal transfer cladding of wear and corrosion resistant coatings in engine applications. Tampere University of Technology, Tampere
18. Liu Y, Liu ZZ, Chen YM et al (2022) Microstructures and properties investigation on DP980 dual-phase steel CMT + P welded joints. Materials 15(17):5880
19. Shu FY, Lü YH, Liu YX et al (2014) FEM modeling of softened base metal in narrow-gap joint by CMT+P MIX welding procedure. Trans Nonferrous Metals Soc China 24(6):1830–1835
20. Wang XR, Wu Y, Pan H et al (2021) Microstructure and softening of advanced high-strength steel QP1180 lap joints welded with CMT. Mater Lett 287:129282
21. Solomon IJ, Sevvel P, Gunasekaran J (2021) A review on the various processing parameters in FDM. Mater Today: Proc 37:509–514
22. Wang DL, Wu CF, Suo YC et al (2019) Effect of pulse frequency on arc behavior and droplet transfer of 2198 Al–Li alloy by ultrahigh-frequency pulse AC CMT welding. J Market Res 8(5):3950–3958

23. Scotti FM, Teixeira FR, da Silva LJ et al (2020) Thermal management in WAAM through the CMT Advanced process and an active cooling technique. J Manuf Process 57:23–35
24. Yang K, Wang F, Duan DS et al (2021) Experimental investigation of integrated circular triple-wire pulse GMAW of Q960E high-strength steel for construction machinery. Materials 14(2):375

MIX
Papier aus verantwortungsvollen Quellen
Paper from responsible sources
FSC® C105338

If you have any concerns about our products,
you can contact us on
ProductSafety@springernature.com

In case Publisher is established outside the EU,
the EU authorized representative is:
Springer Nature Customer Service Center GmbH
Europaplatz 3, 69115 Heidelberg, Germany

Printed by Libri Plureos GmbH
in Hamburg, Germany